General Chemistry I

CHE 121

Slowinski/Wolsey

Australia • Brazil • Japan • Korea • Mexico • Singapore • Spain • United Kingdom • United States

General Chemistry I
CHE 121

Executive Editors:
 Maureen Staudt
 Michael Stranz

Senior Project Development Manager:
 Linda deStefano

Marketing Specialist:
 Courtney Sheldon

Senior Production/Manufacturing Manager:
 Donna M. Brown

PreMedia Manager:
 Joel Brennecke

Sr. Rights Acquisition Account Manager:
 Todd Osborne

Cover Image:
Getty Images*

*Unless otherwise noted, all cover images used by Custom Solutions, a part of Cengage Learning, have been supplied courtesy of Getty Images with the exception of the Earthview cover image, which has been supplied by the National Aeronautics and Space Administration (NASA).

For product information and technology assistance, contact us at
Cengage Learning Customer & Sales Support, 1-800-354-9706

For permission to use material from this text or product,
submit all requests online at **cengage.com/permissions**
Further permissions questions can be emailed to
permissionrequest@cengage.com

This book contains select works from existing Cengage Learning resources and was produced by Cengage Learning Custom Solutions for collegiate use. As such, those adopting and/or contributing to this work are responsible for editorial content accuracy, continuity and completeness.

Compilation © 2011 Cengage Learning
ISBN-13: 978-1-111-76875-1

ISBN-10: 1-111-76875-7

Cengage Learning
5191 Natorp Boulevard
Mason, Ohio 45040
USA

Cengage Learning is a leading provider of customized learning solutions with office locations around the globe, including Singapore, the United Kingdom, Australia, Mexico, Brazil, and Japan. Locate your local office at:
international.cengage.com/region

Cengage Learning products are represented in Canada by Nelson Education, Ltd.
For your lifelong learning solutions, visit **www.cengage.com/custom**
Visit our corporate website at **www.cengage.com**

Printed in the United States of America

Acknowledgements

The content of this text has been adapted from the following product(s):

Chemical Principles in the Laboratory
Slowinski/Wolsey ISBN-10: (0-495-11288-7)
ISBN-13: (978-0-495-11288-4)

Table Of Contents

The Densities of Liquids and Solids

Given a sample of a pure liquid, we can measure many of its characteristics. Its temperature, mass, color, and volume are among the many properties we can determine. We find that, if we measure the mass and volume of different samples of the liquid, the mass and volume of each sample are related in a simple way; if we divide the mass by the volume, the result we obtain is the same for each sample, independent of its mass. That is, for samples A, B, and C, of the liquid at constant temperature and pressure,

$$\text{Mass}_A/\text{Volume}_A = \text{Mass}_B/\text{Volume}_B = \text{Mass}_C/\text{Volume}_C = \text{a constant}$$

That constant, which is clearly independent of the size of the sample, is called its **density**, and is one of the fundamental properties of the liquid. The density of water is exactly 1.00000 g/cm^3 at 4°C, and is slightly less than one at room temperature (0.9970 g/cm^3 at 25°C). Densities of liquids and solids range from values that are less than that of water to values that are much greater. Osmium metal has a density of 22.5 g/cm^3 and is probably the densest material known at ordinary pressures.

In any density determination, two quantities must be determined—the mass and the volume of a given quantity of matter. The mass can easily be determined by weighing a sample of the substance on a balance. The quantity we usually think of as "weight" is really the mass of a substance. In the process of "weighing" we find the mass, taken from a standard set of masses, that experiences the same gravitational force as that experienced by the given quantity of matter we are weighing. The mass of a sample of liquid in a container can be found by taking the difference between the mass of the container plus the liquid and the mass of the empty container.

The volume of a liquid can easily be determined by means of a calibrated container. In the laboratory a graduated cylinder is often used for routine measurements of volume. Accurate measurement of liquid volume is made by using a pycnometer, which is simply a container having a precisely definable volume. The volume of a solid can be determined by direct measurement if the solid has a regular geometrical shape. Such is not usually the case, however, with ordinary solid samples. A convenient way to determine the volume of a solid is to measure accurately the volume of liquid displaced when an amount of the solid is immersed in the liquid. The volume of the solid will equal the volume of liquid which it displaces.

In this experiment we will determine the density of a liquid and a solid by the procedure we have outlined. First we weigh an empty flask and its stopper. We then fill the flask completely with water, measuring the mass of the filled stoppered flask. From the difference in these two masses we find the mass of water and then, from the known density of water, we determine the volume of the flask. We empty and dry the flask, fill it with an unknown liquid, and weigh again. From the mass of the liquid and the volume of the flask we find the density of the liquid. To determine the density of an unknown solid metal, we add the metal to the dry empty flask and weigh. This allows us to find the mass of the metal. We then fill the flask with water, leaving the metal in the flask, and weigh again. The increase in mass is that of the added water; from that increase, and the density of water, we calculate the volume of water we added. The volume of the metal must equal the volume of the flask minus the volume of water. From the mass and volume of the metal we calculate its density. The calculations involved are outlined in detail in the Advance Study Assignment.

WEAR YOUR SAFETY GLASSES WHILE
PERFORMING THIS EXPERIMENT

Experimental Procedure

A. Mass of a Coin

After you have been shown how to operate the analytical balances in your laboratory, read the section on balances in Appendix IV. Take a coin and measure its mass to 0.0001 g. Record the mass on the Data page. If your balance has a TARE bar, use it to re-zero the balance. Take another coin and weigh it, recording its mass. Remove both coins, zero the balance, and weigh both coins together, recording the total mass. If you have no TARE bar on your balance, add the second coin and measure and record the mass of the two coins. Then remove both coins and find the mass of the second one by itself. When you are satisfied that your results are those you would expect, go to the stockroom and obtain a glass-stoppered flask, which will serve as a pycnometer, and samples of an unknown liquid and an unknown metal.

B. Density of a Liquid

If your flask is not clean and dry, clean it with detergent solution and water, rinse it with a few cubic centimeters of acetone, and dry it by letting it stand for a few minutes in the air or by *gently* blowing compressed air into it for a few moments.

Weigh the dry flask with its stopper on the analytical balance, or the toploading balance if so directed, to the nearest milligram. Fill the flask with distilled water until the liquid level is nearly to the *top* of the ground surface in the neck. Put the stopper in the flask in order to drive out *all* the air and any excess water. Work the stopper gently into the flask, so that it is firmly seated in position. Wipe any water from the outside of the flask with a towel and soak up all excess water from around the top of the stopper.

Again weigh the flask, which should be completely dry on the outside and full of water, to the nearest milligram. Given the density of water at the temperature of the laboratory and the mass of water in the flask, you should be able to determine the volume of the flask very precisely. Empty the flask, dry it, and fill it with your unknown liquid. Stopper and dry the flask as you did when working with the water, and then weigh the stoppered flask full of the unknown liquid, making sure its surface is dry. This measurement, used in conjunction with those you made previously, will allow you to find accurately the density of your unknown liquid.

C. Density of a Solid

Pour your sample of liquid from the flask into its container. Rinse the flask with a small amount of acetone and dry it thoroughly. Add small chunks of the metal sample to the flask until the flask is at least half full. Weigh the flask, with its stopper and the metal, to the nearest milligram. You should have at least 50 g of metal in the flask.

Leaving the metal in the flask, fill the flask with water and then replace the stopper. Roll the metal around in the flask to make sure that no air remains between the metal pieces. Refill the flask if necessary, and then weigh the dry, stoppered flask full of water plus the metal sample. Properly done, the measurements you have made in this experiment will allow a calculation of the density of your metal sample that will be accurate to about 0.1%.

DISPOSAL OF REACTION PRODUCTS. Pour the water from the flask. Put the metal in its container. Dry the flask and return it with its stopper and your metal sample, along with the sample of unknown liquid, to the stockroom.

Experiment 1

Data and Calculations: The Densities of Liquids and Solids

A. Mass of coin 1 _____ g   Mass of coin 2 _____ g

Mass of coins 1 and 2 weighed together _____ g
What general law is illustrated by the results of this experiment?

B. Density of unknown liquid

Mass of empty flask plus stopper _____ g

Mass of stoppered flask plus water _____ g

Mass of stoppered flask plus liquid _____ g

Mass of water _____ g

Temperature in the laboratory _____ °C

Volume of flask (density of H_2O at 25°C, 0.9970 g/cm³; at 20°C, 0.9982 g/cm³) _____ cm³

Mass of liquid _____ g

Density of liquid _____ g/cm³

To how many significant figures can the liquid density be properly reported? (See Appendix V.) _____

C. Density of unknown metal

Mass of stoppered flask plus metal _____ g

Mass of stoppered flask plus metal plus water _____ g

Mass of metal _____ g

Mass of water _____ g

Volume of water _____ cm³

Volume of metal   _____ cm^3

Density of metal   _____ g/cm^3

To how many significant figures can the density of the metal be
properly reported?   _____

Explain why the value obtained for the density of the metal is likely to have a larger percentage error than
that found for the liquid.

Unknown liquid no. _____   Unknown solid no. _____

Experiment 1

Advance Study Assignment: Densities of Solids and Liquids

The advance study assignments in this laboratory manual are designed to assist you in making the calculations required in the experiment you will be doing. We do this by furnishing you with sample data and showing in some detail how that data can be used to obtain the desired results. In the advance study assignments we will often include the guiding principles as well as the specific relationships to be employed. If you work through the steps in each calculation by yourself, you should have no difficulty when you are called upon to make the necessary calculations on the basis of the data you obtain in the laboratory.

1. **Finding the volume of a flask.**

    A student obtained a clean, dry glass-stoppered flask. She weighed the flask and stopper on an analytical balance and found the total mass to be 32.634 g. She then filled the flask with water and obtained a mass for the full stoppered flask of 59.479 g. From these data, and the fact that at the temperature of the laboratory the density of water was 0.9973 g/cm^3, find the volume of the stoppered flask.

    a. First we need to obtain the mass of the water in the flask. This is found by recognizing that the mass of a sample is equal to the sum of the masses of its parts. For the filled stoppered flask:

    Mass of filled stoppered flask = mass of empty stoppered flask + mass of water,
        so mass of water = mass of filled flask – mass of empty flask

    Mass of water = _____ g – _____ g = _____ g

    Many mass and volume measurements in chemistry are made by the method used in 1a. This method is called measuring by difference, and is a very useful one.

    b. The density of a pure substance is equal to its mass divided by its volume:

    $$\text{Density} = \frac{\text{mass}}{\text{volume}} \quad \text{or} \quad \text{volume} = \frac{\text{mass}}{\text{density}}$$

    The volume of the flask is equal to the volume of the water it contains. Since we know the mass and density of the water, we can find its volume and that of the flask. Make the necessary calculation.

    Volume of water = volume of flask = _____ cm^3

2. **Finding the density of an unknown liquid.**

    Having obtained the volume of the flask, the student emptied the flask, dried it, and filled it with an unknown whose density she wished to determine. The mass of the stoppered flask when completely filled with liquid was 50.376 g. Find the density of the liquid.

    a. First we need to find the mass of the liquid by measuring by difference:

    Mass of liquid = _____ g – _____ g = _____ g

b. Since the volume of the liquid equals that of the flask, we know both the mass and volume of the liquid and can easily find its density using the equation in 1b. Make the calculation.

Density of liquid = _____ g/cm^3

3. Finding the density of a solid.

The student then emptied the flask and dried it once again. To the empty flask she added pieces of a metal until the flask was about half full. She weighed the stoppered flask and its metal contents and found that the mass was 152.047 g. She then filled the flask with water, stoppered it, and obtained a total mass of 165.541 g for the flask, stopper, metal, and water. Find the density of the metal.

a. To find the density of the metal we need to know its mass and volume. We can easily obtain its mass by the method of differences:

Mass of metal = _____ g – _____ g = _____ g

b. To determine the volume of metal, we note that the volume of the flask must equal the volume of the metal plus the volume of water in the filled flask containing both metal and water. If we can find the volume of water, we can obtain the volume of metal by the method of differences. To obtain the volume of the water we first calculate its mass:

Mass of water = mass of (flask + stopper + metal + water) – mass of (flask + stopper + metal)

Mass of water = _____ g – _____ g = _____ g

The volume of water is found from its density, as in 1b. Make the calculation.

Volume of water = _____ cm^3

c. From the volume of the water, we calculate the volume of metal:

Volume of metal = volume of flask – volume of water

Volume of metal = _____ cm^3 – _____ cm^3 = _____ cm^3

From the mass of and volume of metal, we find the density, using the equation in 1b. Make the calculation.

Density of metal = _____ g/cm^3

Now go back to Question 1 and check to see that you have reported the proper number of significant figures in each of the results you calculated in this assignment. Use the rules on significant figures as given in your chemistry text or Appendix V.

Resolution of Matter into Pure Substances,
I. Paper Chromatography

The fact that different substances have different solubilities in a given solvent can be used in several ways to effect a separation of substances from mixtures in which they are present. We will see in an upcoming experiment how fractional crystallization allows us to obtain pure substances by relatively simple procedures based on solubility properties. Another widely used resolution technique, which also depends on solubility differences, is chromatography.

In the chromatographic experiment a mixture is deposited on some solid adsorbing substance, which might consist of a strip of filter paper, a thin layer of silica gel on a piece of glass, some finely divided charcoal packed loosely in a glass tube, or even some microscopic glass beads coated thinly with a suitable adsorbing substance and contained in a piece of copper tubing.

The components of a mixture are adsorbed on the solid to varying degrees, depending on the nature of the component, the nature of the adsorbent, and the temperature. A solvent is then caused to flow through the adsorbent solid under applied or gravitational pressure or by the capillary effect. As the solvent passes the deposited sample, the various components tend, to varying extents, to be dissolved and swept along the solid. The rate at which a component will move along the solid depends on its relative tendency to be dissolved in the solvent and adsorbed on the solid. The net effect is that, as the solvent passes slowly through the solid, the components separate from each other and move along as rather diffuse zones. With the proper choice of solvent and adsorbent, it is possible to resolve many complex mixtures by this procedure. If necessary, we can usually recover a given component by identifying the position of the zone containing the component, removing that part of the solid from the system, and eluting the desired component with a suitable good solvent.

The name given to a particular kind of chromatography depends upon the manner in which the experiment is conducted. Thus, we have column, thin-layer, paper, and gas chromatography, all in very common use (Fig. 2.1). Chromatography in its many possible variations offers the chemist one of the best methods, if not the best method, for resolving a mixture into pure substances, regardless of whether that mixture consists of a gas, a volatile liquid, or a group of nonvolatile, relatively unstable, complex organic compounds.

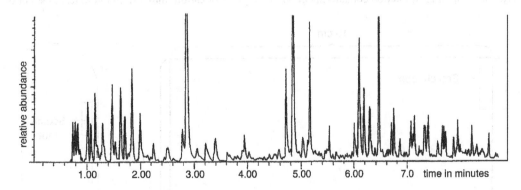

Figure 2.1   This is a gas chromatogram of a sample of unleaded gasoline. Each peak corresponds to a different molecule, so gasoline has many components, at least 50, each of which can be identified. The molar masses vary from about 50 to about 150, with the largest peak, at about 3 min, that due to toluene, $C_6H_5CH_3$. Sample size for the chromatogram was less than 10^{-6} grams, <0.001 mg! Gas chromatography offers the best method for resolution of complex volatile mixtures. Chromatogram courtesy of Prof. Becky Hoye at Macalester College.

In this experiment we will use paper chromatography to separate a mixture of metallic ions in solution. A sample containing a few micrograms of ions is applied as a spot near one edge of a piece of filter paper. That edge is immersed in a solvent, with the paper held vertically. As the solvent rises up the paper by capillary action, it will carry the metallic ions along with it to a degree that depends upon the relative tendency of each ion to dissolve in the solvent and adsorb on the paper. Because the ions differ in their properties, they move at different rates and become separated on the paper. The position of each ion during the experiment can be recognized if the ion is colored, as some of them are. At the end of the experiment their positions are established more clearly by treating the paper with a staining reagent which reacts with each ion to produce a colored product. By observing the position and color of the spot produced by each ion, and the positions of the spots produced by an unknown containing some of those ions, you can readily determine the ions present in the unknown.

It is possible to describe the position of spots such as those you will be observing in terms of a quantity called the R_f value. In the experiment the solvent rises a certain distance, say L centimeters. At the same time a given component will usually rise a smaller distance, say D centimeters. The ratio of D/L is called the R_f value for that component:

$$R_f = \frac{D}{L} = \frac{\text{distance component moves}}{\text{distance solvent moves}} \qquad (1)$$

The R_f value is a characteristic property of a given component in a chromatography experiment conducted under particular conditions. It does not depend upon concentration or upon the other components present. Hence it can be reported in the literature and used by other researchers doing similar analyses. In the experiment you will be doing, you will be asked to calculate the R_f values for each of the cations studied.

Experimental Procedure

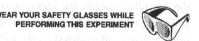

WEAR YOUR SAFETY GLASSES WHILE
PERFORMING THIS EXPERIMENT

From the stockroom obtain an unknown and a piece of filter paper about 19 cm long and 11 cm wide. Along the 19-cm edge, draw a pencil line about 1 cm from that edge. Starting 1.5 cm from the end of the line, mark the line at 2-cm intervals. Label the segments of the line as shown in Figure 2.2, with the formulas of the ions to be studied and the known and unknown mixtures.

Put two or three drops of 0.1 M solutions of the following compounds in small micro test tubes, one solution to a tube:

$$AgNO_3 \quad Co(NO_3)_2 \quad Cu(NO_3)_2 \quad Fe(NO_3)_3 \quad Hg(NO_3)_2$$

In solution these substances exist as ions. The metallic cations are Ag^+, Co^{2+}, Cu^{2+}, Fe^{3+}, and Hg^{2+}, respectively. One drop of each solution contains about 50 micrograms of cation. Into a sixth micro test tube put two

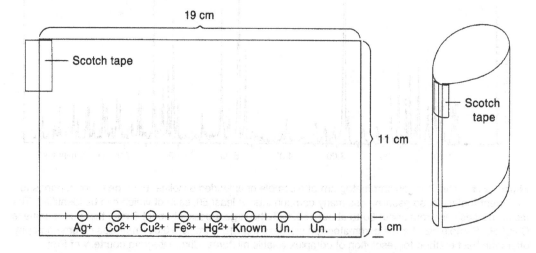

drops of each of the five solutions; swirl until the solutions are well mixed. This mixture will be our known, since we know it contains all of the cations.

Your instructor will furnish you with a fine capillary tube, which will serve as an applicator. Test the application procedure by dipping the applicator into one of the colored solutions and touching it momentarily to a round piece of filter paper. The liquid from the applicator should form a spot no larger than 8 mm in diameter. Practice making spots until you can reproduce the spot size each time.

Clean the applicator by dipping it about 1 cm into distilled water and then touching the round filter paper to remove the liquid. Continue contact until all the liquid in the tube is gone. Repeat the cleaning procedure one more time. Dip the applicator into one of the cation solutions and put a spot on the line on the rectangular filter paper in the region labeled for that cation. Clean the applicator twice, and repeat the procedure with another solution. Continue this approach until you have put a spot for each of the five cations and the known and unknown on the paper, cleaning the applicator between solutions. Dry the paper by moving it in the air or holding it briefly in front of a hair dryer or heat lamp (low setting). Apply the known and unknown three more times to the same spots; the known and unknown are less concentrated than the cation solutions, so this procedure will increase the amount of each ion in the spots. Make sure that you dry the spots between applications, since otherwise they will get larger. Don't heat the paper more than necessary, just enough to dry the spots.

Draw about 15 mL of eluting solution from the supply on the reagent shelf. This solution is made by mixing a solution of HCl, hydrochloric acid, with ethanol and butanol, which are organic solvents. Pour the eluting solution into a 600-mL beaker and cover with a watch glass.

Check to make sure that the spots on the filter paper are all dry. Place a 4- to 5-cm length of Scotch tape along the upper end of the left edge of the paper, as shown in Figure 2.2, so that about half of the tape is on the paper. Form the paper into a cylinder by attaching the tape to the other edge, in such a way that the edges are parallel but do not overlap. When you are finished, the pencil line at the bottom of the cylinder should form a circle, approximately anyway, and the two edges of the paper should not quite touch. Stand the cylinder up on the lab bench to check that such is the case and readjust the tape if necessary. *Do not* tape the lower edges of the paper together.

Place the cylinder in the eluting solution in the 600-mL beaker, with the sample spots down near the liquid surface. The paper should not touch the wall of the beaker. Cover the beaker with the watch glass. The solvent will gradually rise by capillary action up the filter paper, carrying along the cations at different rates. After the process has gone on for a few minutes, you should be able to see colored spots on the paper, showing the positions of some of the cations.

While the experiment is proceeding, you can test the effect of the staining reagent on the different cations. Put an 8-mm spot of each of the cation solutions on a clean piece of round filter paper, labeling each spot and cleaning the applicator between solutions. Dry the spots as before. Some of them will have a little color; record those colors on the Data page. Put the filter paper on a paper towel, and, using the spray bottle on the lab bench, spray the paper evenly with the staining reagent, getting the paper moist but not really wet. The staining reagent is a solution containing potassium ferrocyanide and potassium iodide. This reagent forms colored precipitates or reaction products with many cations, including all of those used in this experiment. Note the colors obtained with each of the cations. Considering that each spot contains less than 50 micrograms of cation, the tests are quite definitive.

When the eluting solution has risen to within about 2 cm of the top of the filter paper (it will take about 75 minutes), remove the cylinder from the beaker and take off the tape. Draw a pencil line along the solvent front. Dry the paper with gentle heat until it is quite dry. Note any cations that must be in your unknown by virtue of your being able to see their colors. Then, with the paper on a paper towel, spray it as before with the staining reagent. Any cations you identified in your unknown before staining should be observed, as well as any that require staining for detection.

Measure the distance from the straight line on which you applied the spots to the solvent front, which is distance L in Equation 1. Then measure the distance from the pencil line to the center of the spot made by each of the cations, when pure and in the known; this is distance D. Calculate the R_f value for each cation. Then calculate R_f values for the cations in the unknown. How do the R_f values compare?

DISPOSAL OF REACTION PRODUCTS.   When you are finished with the experiment, pour the eluting solution into the waste crock, not down the sink. Wash your hands before leaving the laboratory.

Experiment 2

Data and Calculations: Resolution of Matter into Pure Substances, I. Paper Chromatography

	Ag^+	Co^{2+}	Cu^{2+}	Fe^{3+}	Hg^{2+}
Colors (if observed)					
Dry					
After staining					
Distance solvent moved (L)					
Distance cation moved (D)					
R_f					

Known Mixture

Distance solvent moved					
Distance cation moved					
R_f					

Unknown Mixture

Cations identified _____

Dry					
After staining					
Distance solvent moved					
Distance cation moved					
R_f					

Composition of unknown _____

Unknown no. _____

Experiment 2

Advance Study Assignment: Resolution of Matter into Pure Substances, I. Paper Chromatography

1. A student chromatographs a mixture, and after developing the spots with a suitable reagent he observes the following:

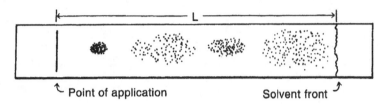

What are the R_f values?

2. Explain, in your own words, why samples can often be separated into their components by chromatography.

3. The solvent moves 3 cm in about 10 minutes. Why shouldn't the experiment be stopped at that time instead of waiting 75 minutes for the solvent to move 10 cm?

4. In this experiment it takes about 10 microliters of solution to produce a spot 1 cm in diameter. If the $Cu(NO_3)_2$ solution contains about 6 g Cu^{2+} per liter, how many micrograms of Cu^{2+} ion are there in one spot?

3

Determination of a Chemical Formula

When atoms of one element combine with those of another, the combining ratio is typically an integer or a simple fraction; 1:2, 1:1, 2:1, and 2:3 are ratios one might encounter. The simplest formula of a compound expresses that atom ratio. Some substances with the ratios we listed include $CaCl_2$, KBr, Ag_2O, and Fe_2O_3. When more than two elements are present in a compound, the formula still indicates the atom ratio. Thus the substance with the formula Na_2SO_4 indicates that the sodium, sulfur, and oxygen atoms occur in that compound in the ratio 2:1:4. Many compounds have more complex formulas than those we have noted, but the same principles apply.

To find the formula of a compound we need to find the mass of each of the elements in a weighed sample of that compound. For example, if we resolved a sample of the compound NaOH weighing 40 grams into its elements, we would find that we obtained just about 23 grams of sodium, 16 grams of oxygen, and 1 gram of hydrogen. Since the atomic mass scale tells us that sodium atoms have a relative mass of 23, oxygen atoms a relative mass of 16, and hydrogen atoms a relative mass of just about 1, we would conclude that the sample of NaOH contained equal numbers of Na, O, and H atoms. Since that is the case, the atom ratio Na:O:H is 1:1:1, and so the simplest formula is NaOH. In terms of moles, we can say that that one mole of NaOH, 40 grams, contains one mole of Na, 23 grams, one mole of O, 16 grams, and one mole of H, 1 gram, where we define the mole to be that mass in grams equal numerically to the sum of the atomic masses in an element or a compound. From this kind of argument we can conclude that the atom ratio in a compound is equal to the mole ratio. We get the mole ratio from chemical analysis, and from that the formula of the compound.

In this experiment we will use these principles to find the formula of the compound with the general formula $Cu_xCl_y \cdot zH_2O$, where the x, y, and z are integers which, when known, establish the formula of the compound. (In expressing the formula of a compound like this one, where water molecules remain intact within the compound, we retain the formula of H_2O in the formula of the compound.)

The compound we will study, which is called copper chloride hydrate, turns out to be ideal for one's first venture into formula determination. It is stable, can be obtained in pure form, has a characteristic blue-green color which changes as the compound is changed chemically, and is relatively easy to decompose into the elements and water. In the experiment we will first drive out the water, which is called the water of hydration, from an accurately weighed sample of the compound. This occurs if we gently heat the sample to a little over 100°C. As the water is driven out, the color of the sample changes from blue-green to a tan-brown color similar to that of tobacco. The compound formed is anhydrous (no water) copper chloride. If we subtract its mass from that of the hydrate, we can determine the mass of the water that was driven off, and, using the molar mass of water, find the number of moles of H_2O that were in the sample.

In the next step we need to find either the mass of copper or the mass of chlorine in the anhydrous sample we have prepared. It turns out to be much easier to determine the mass of the copper, and find the mass of chlorine by difference. We do this by dissolving the anhydrous sample in water, which gives us a green solution containing copper and chloride ions. To that solution we add some aluminum metal wire. Aluminum is what we call an active metal; in contact with a solution containing copper ions, the aluminum metal will react chemically with those ions, converting them to copper metal. The aluminum is said to reduce the copper ions to the metal, and is itself oxidized. The copper metal appears on the wire as the reaction proceeds, and has the typical red-orange color. When the reaction is complete, we remove the excess Al, separate the copper from the solution, and weigh the dried metal. From its mass we can calculate the number of moles of copper in the sample. We find the mass of chlorine by subtracting the mass of copper from that of the anhydrous copper chloride, and from that value determine the number of moles of chlorine. The mole ratio for Cu:Cl:H_2O gives us the formula of the compound.

Experimental Procedure

Weigh a clean, dry crucible, without a cover, accurately on the analytical balance. Place about 1 gram of the unknown hydrated copper chloride in the crucible. With your spatula, break up any sizeable crystal particles by pressing them against the wall of the crucible. Then weigh the crucible and its contents accurately. Enter your results on the Data page.

Place the uncovered crucible on a clay triangle supported by an iron ring. Light your Bunsen burner away from the crucible, and adjust the burner so that you have a small flame. Holding the burner in your hand, gently heat the crucible as you move the burner back and forth. Do not overheat the sample. As the sample warms, you will see that the green crystals begin to change to brown around the edges. Continue gentle heating, slowly converting all of the hydrated crystals to the anhydrous brown form. After all of the crystals appear to be brown, continue heating gently, moving the burner back and forth judiciously, for an additional two minutes. Remove the burner, cover the crucible to minimize rehydration, and let it cool for about 15 minutes. Remove the cover, and slowly roll the brown crystals around the crucible. If some green crystals remain, repeat the heating process. Finally, weigh the cool uncovered crucible and its contents accurately.

Transfer the brown crystals in the crucible to an empty 50-mL beaker. Rinse out the crucible with two 5- to 7-mL portions of distilled water, and add the rinsings to the beaker. Swirl the beaker gently to dissolve the brown solid. The color will change to green as the copper ions are rehydrated. Measure out about 20 cm of 20-gauge aluminum wire (~0.25 g) and form the wire into a loose spiral coil. Put the coil into the solution so that it is completely immersed. Within a few moments you will observe some evolution of H_2, hydrogen gas, and the formation of copper metal on the Al wire. As the copper ions are reduced, the color of the solution will fade. The Al metal wire will be slowly oxidized and enter the solution as aluminum ions. (The hydrogen gas is formed as the aluminum reduces water in the slightly acidic copper solution.)

When the reaction is complete, which will take about 30 minutes, the solution will be colorless, and most of the copper metal that was produced will be on the Al wire. Add 5 drops of 6 M HCl to dissolve any insoluble aluminum salts and clear up the solution. Use your glass stirring rod to remove the copper from the wire as completely as you can. Slide the unreacted aluminum wire up the wall of the beaker with your stirring rod, and, while the wire is hanging from the rod, rinse off any remaining Cu particles with water from your wash bottle. If necessary, complete the removal of the Cu with a drop or two of 6 M HCl added directly to the wire. Put the wire aside; it has done its duty.

In the beaker you now have the metallic copper produced in the reaction, in a solution containing an aluminum salt. In the next step we will use a Buchner funnel to separate the copper from the solution. Weigh accurately a dry piece of filter paper that will fit in the Buchner funnel, and record its mass. Put the paper in the funnel, and apply light suction as you add a few mL of water to ensure a good seal. With suction on, decant the solution into the funnel. Wash the copper metal thoroughly with distilled water, breaking up any copper particles with your stirring rod. Transfer the wash and the copper to the filter funnel. Wash any remaining copper into the funnel with water from your wash bottle. **All** of the copper must be transferred to the funnel. Rinse the copper on the paper once again with water. Turn off the suction. Add 10 mL of 95% ethanol to the funnel, and after a minute or so turn on the suction. Draw air through the funnel for about 5 minutes. With your spatula, lift the edge of the paper, and carefully lift the paper and the copper from the funnel. Dry the paper and copper under a heat lamp for 5 minutes. Allow it to cool to room temperature and then weigh it accurately.

DISPOSAL OF REACTION PRODUCTS. Dispose of the liquid waste and copper produced in the experiment as directed by your instructor.

Experiment 4

Data and Calculations: Determination of a Chemical Formula

Atomic masses: Copper _____ Cl _____ H _____ O _____

Mass of crucible   _____ g

Mass of crucible and hydrated sample   _____ g

Mass of hydrated sample   _____ g

Mass of crucible and dehydrated sample   _____ g

Mass of dehydrated sample   _____ g

Mass of filter paper   _____ g

Mass of filter paper and copper   _____ g

Mass of copper   _____ g

No. moles of copper   _____ moles

Mass of water evolved   _____ g

No. moles of water   _____ moles

Mass of chlorine in sample (by difference)   _____ g

No. moles of chlorine   _____ moles

Mole ratio, chlorine:copper in sample   _____ :1

Mole ratio, water:copper in hydrated sample   _____ :1

Formula of dehydrated sample (round to nearest integer)   _____

Formula of hydrated sample   _____

Experiment 4

Advance Study Assignment: Determination of a Chemical Formula

1. To find the mass of a mole of an element, one looks up the atomic mass of the element in a table of atomic masses (see Appendix III or the Periodic Table). The molar mass of an element is simply the mass in grams of that element that is numerically equal to its atomic mass. For a compound substance, the molar mass is equal to the mass in grams that is numerically equal to the sum of the atomic masses in the formula of the substance. Find the molar mass of

    Cu _____ g   Cl _____ g   H _____ g   O _____ g H_2O _____ g

2. If one can find the ratio of the number of moles of the elements in a compound to one another, one can find the formula of the compound. In a certain compound of copper and oxygen, Cu_xO_y, we find that a sample weighing 0.6349 g contains 0.5072 g Cu.

    a. How many moles of Cu are there in the sample?

    $$\left(No. \ moles = \frac{mass \ Cu}{molar \ mass \ Cu} \right)$$

    _____ moles

    b. How many grams of O are there in the sample? (The mass of the sample equals the mass of Cu plus the mass of O.)

    _____ g

    c. How many moles of O are there in the sample?

    _____ moles

    d. What is the mole ratio (no. moles Cu/no. moles O) in the sample?

    _____ : 1

    e. What is the formula of the oxide? (The atom ratio equals the mole ratio, and is expressed using the smallest integers possible.)

    _____

    f. What is the molar mass of the copper oxide?

    _____ g

4

Properties of Hydrates

Most solid chemical compounds will contain some water if they have been exposed to the atmosphere for any length of time. In most cases the water is present in very small amounts, and is merely adsorbed on the surface of the crystals. Other solid compounds contain larger amounts of water that is chemically bound in the crystal. These compounds are usually ionic salts. The water that is present in these salts is called water of hydration and is usually bound to the cations in the salt.

The water molecules in a hydrate are removed relatively easily. In many cases, simply heating a hydrate to a temperature somewhat above the boiling point of water will drive off the water of hydration. Hydrated barium chloride is typical in this regard; it is converted to anhydrous $BaCl_2$ if heated to about 115°C:

$$BaCl_2 \cdot 2\,H_2O(s) \rightarrow BaCl_2(s) + 2\,H_2O(g) \text{ at } t \geq 115°C$$

In the dehydration reaction the crystal structure of the solid will change, and the color of the salt may also change. In Experiment 4, when copper chloride hydrate was gently heated, it was converted to the brownish anhydride.

Some hydrates lose water to the atmosphere upon standing in air. This process is called efflorescence. The amount of water lost depends on the amount of water in the air, as measured by its relative humidity, and the temperature. In moist warm air, $CoCl_2$ is fully hydrated and exists as $CoCl_2 \cdot 6\,H_2O$, which is red. In cold dry air the salt loses most of its water of hydration and is found as anhydrous $CoCl_2$, which is blue. At intermediate humidities and 25°C, we find the stable form is the violet dihydrate, $CoCl_2 \cdot 2\,H_2O$. In the old days one could obtain inexpensive hygrometers that indicated the humidity by the color of the cobalt chloride they contained.

Some anhydrous ionic compounds will tend to absorb water from the air or other sources so strongly that they can be used to dry liquids or gases. These substances are called desiccants, and are said to be hygroscopic. A few ionic compounds can take up so much water from the air that they dissolve in the water they absorb; sodium hydroxide, NaOH, will do this. This process is called deliquescence.

Some compounds evolve water on being heated but are not true hydrates. The water is produced by decomposition of the compound rather than by loss of water of hydration. Organic compounds, particularly carbohydrates, behave this way. Decompositions of this sort are not reversible; adding water to the product will not regenerate the original compound. True hydrates typically undergo reversible dehydration. Adding water to anhydrous $BaCl_2$ will cause formation of $BaCl_2 \cdot 2\,H_2O$, or if enough water is added you will get a solution containing Ba^{2+} and Cl^- ions. Many ionic hydrates are soluble in water, and are usually prepared by crystallization from water solution. If you order barium chloride from a chemical supply house, you will probably get crystals of $BaCl_2 \cdot 2\,H_2O$, which is a stable, stoichiometrically pure, compound. The amount of bound water may depend on the way the hydrate is prepared, but in general the number of moles of water per mole of compound is either an integer or a multiple of ½.

In this experiment you will study some of the properties of hydrates. You will identify the hydrates in a group of compounds, observe the reversibility of the hydration reaction, and test some substances for efflorescence or deliquescence. Finally you will be asked to determine the amount of water lost by a sample of unknown hydrate on heating. From this amount, if given the formula or the molar mass of the anhydrous sample, you will be able to calculate the formula of the hydrate itself.

Experimental Procedure

WEAR YOUR SAFETY GLASSES WHILE
PERFORMING THIS EXPERIMENT

A. Identification of Hydrates

Place about 0.5 g of each of the compounds listed below in small, dry test tubes, one compound to a tube. Observe carefully the behavior of each compound when you heat it gently with a burner flame. If droplets of water condense on the cool upper walls of the test tube, this is evidence that the compound may be a hydrate. Note the nature and color of the residue. Let the tube cool and try to dissolve the residue in a few cm³ of water, warming very gently if necessary. A true hydrate will tend to dissolve in water, producing a solution with a color very similar to that of the original hydrate. If the compound is a carbohydrate, it will give off water on heating and will tend to char. The solution of the residue in water will often be caramel colored.

Nickel chloride	Sucrose
Potassium chloride	Calcium carbonate
Sodium tetraborate (borax)	Barium chloride

B. Reversibility of Hydration

Gently heat a few crystals, ~0.3 g, of hydrated cobalt(II) chloride, $CoCl_2 \cdot 6 H_2O$, in an evaporating dish until the color change appears to be complete. Dissolve the residue in the evaporating dish in a few cm³ of water from your wash bottle. Heat the resulting solution to boiling **CAUTION:** and carefully boil it to dryness. Note any color changes. Put the evaporating dish on the lab bench and let it cool.

C. Deliquescence and Efflorescence

Place a few crystals of each of the compounds listed below on separate watch glasses and put them next to the dish of $CoCl_2$ prepared in Part B. Depending on their composition and the relative humidity (amount of moisture in the air), the samples may gradually lose water of hydration to, or pick up water from, the air. They may also remain unaffected. To establish whether the samples gain or lose mass, weigh each of them on a top-loading balance to 0.01 g. Record their masses. Weigh them again after about an hour to detect any change in mass. Observe the samples occasionally during the laboratory period, noting any changes in color, crystal structure, or degree of wetness that may occur.

$Na_2CO_3 \cdot 10 H_2O$ (washing soda)	$KAl(SO_4)_2 \cdot 12 H_2O$ (alum)
$CaCl_2$	$CuSO_4$

D. Percent Water in a Hydrate

Clean a porcelain crucible and its cover with 6 M HNO_3. Any stains that are not removed by this treatment will not interfere with this experiment. Rinse the crucible and cover with distilled water. Put the crucible with its cover slightly ajar on a clay triangle and heat with a burner flame, gently at first and then to redness for about 2 minutes. Allow the crucible and cover to cool, and then weigh them to 0.001 g on an analytical balance. Handle the crucible with clean crucible tongs.

Obtain a sample of unknown hydrate from the stockroom and place about a gram of sample in the crucible. Weigh the crucible, cover, and sample on the balance. Put the crucible on the clay triangle, with the cover in an off-center position to allow the escape of water vapor. Heat again, gently at first and then strongly, keeping the bottom of the crucible at red heat for about 10 minutes. Center the cover on the crucible and let it cool to room temperature. Weigh the cooled crucible along with its cover and contents.

Examine the solid residue. Add water until the crucible is two thirds full and stir. Warm gently if the residue does not dissolve readily. Does the residue appear to be soluble in water?

DISPOSAL OF REACTION PRODUCTS.   Dispose of the residues in this experiment as directed by your instructor.

Experiment 6

Data and Calculations: Properties of Hydrates

A. Identification of Hydrates

	H_2O appears	Color of residue	Water soluble	Hydrate
Nickel chloride	_____	_____	_____	_____
Potassium chloride	_____	_____	_____	_____
Sodium tetraborate	_____	_____	_____	_____
Sucrose	_____	_____	_____	_____
Calcium carbonate	_____	_____	_____	_____
Barium chloride	_____	_____	_____	_____

B. Reversibility of Hydration

Summarize your observations on $CoCl_2 \cdot 6 H_2O$:

Is the dehydration and hydration of $CoCl_2$ reversible?

C. Deliquescence and Efflorescence

	Mass (sample + glass)		Observations	Conclusions
	Initial	Final		
$Na_2CO_3 \cdot 10 H_2O$	_____	_____	_____	_____
$KAl(SO_4)_2 \cdot 12 H_2O$	_____	_____	_____	_____
$CaCl_2$	_____	_____	_____	_____
$CuSO_4$	_____	_____	_____	_____
$CoCl_2$	_____	_____	_____	_____

D. Percent Water in a Hydrate

Mass of crucible and cover   _____ g

Mass of crucible, cover, and solid hydrate   _____ g

Mass of crucible, cover, and residue   _____ g

Calculations and Results

Mass of solid hydrate   _____ g

Mass of residue   _____ g

Mass of H_2O lost   _____ g

Percentage of H_2O in the unknown hydrate   _____ %

Formula mass of anhydrous salt (if furnished)   _____

Number of moles of water per mole of unknown hydrate   _____

Unknown no.   _____

Experiment 6

Advance Study Assignment: Properties of Hydrates

1. A student is given a sample of a green nickel sulfate hydrate. She weighs the sample in a dry covered crucible and obtains a mass of 22.326 g for the crucible, cover, and sample. Earlier she had found that the crucible and cover weighed 21.244 g. She then heats the crucible to drive off the water of hydration, keeping the crucible at red heat for about 10 minutes with the cover slightly ajar. She then lets the crucible cool, and finds it has a lower mass; the crucible, cover and contents then weigh 21.840 g. In the process the sample was converted to yellow anhydrous $NiSO_4$.

    a. What was the mass of the hydrate sample?

                                                  _____ g hydrate

    b. What is the mass of the anhydrous $NiSO_4$?

                                                  _____ g $NiSO_4$

    c. How much water was driven off?

                                                  _____ g H_2O

    d. What is the percentage of water in the hydrate?

$$\% \text{ water} = \frac{\text{mass of water in sample}}{\text{mass of hydrate sample}} \times 100\%$$

                                                  _____ % H_2O

    e. How many grams of water would there be in 100.0 g hydrate? How many moles?

                            _____ g H_2O; _____ moles H_2O

    f. How many grams of $NiSO_4$ are there in 100.0 g hydrate? How many moles? (What percentage of the hydrate is $NiSO_4$? Convert the mass of $NiSO_4$ to moles. Molar mass of $NiSO_4$ is 154.8 g.)

                            _____ g $NiSO_4$; _____ moles $NiSO_4$

    g. How many moles of water are present per mole $NiSO_4$?

                                                  _____

    h. What is the formula of the hydrate?

                                                   _____

5

Analysis of an Unknown Chloride

One of the important applications of precipitation reactions lies in the area of quantitative analysis. Many substances that can be precipitated from solution are so slightly soluble that the precipitation reaction by which they are formed can be considered to proceed to completion. Silver chloride is an example of such a substance. If a solution containing Ag^+ ion is slowly added to one containing Cl^- ion, the ions will react to form AgCl:

$$Ag^+(aq) + Cl^-(aq) \rightarrow AgCl(s) \tag{1}$$

Silver chloride is so insoluble that essentially all of the Ag^+ added will precipitate as AgCl until all of the Cl^- is used up. When the amount of Ag^+ added to the solution is equal to the amount of Cl^- initially present, the precipitation of Cl^- ion will be, for all practical purposes, complete.*

A convenient method for chloride analysis using AgCl has been devised. A solution of $AgNO_3$ is added to a chloride solution just to the point where the number of moles of Ag^+ added is equal to the number of moles of Cl^- initially present. We analyze for Cl^- by simply measuring how many moles of $AgNO_3$ are required. Surprisingly enough, this measurement is rather easily made by an experimental procedure called a titration.

In the titration a solution of $AgNO_3$ of known concentration (in moles $AgNO_3$ per liter of solution) is added from a calibrated buret to a solution containing a measured amount of unknown. The titration is stopped when a color change occurs in the solution, indicating that stoichiometrically equivalent amounts of Ag^+ and Cl^- are present. The color change is caused by a chemical reagent, called an indicator, that is added to the solution at the beginning of the titration.

The volume of $AgNO_3$ solution that has been added up to the time of the color change can be measured accurately with the buret, and the number of moles of Ag^+ added can be calculated from the known concentration of the solution.

In the Mohr method for the volumetric analysis of chloride, which we will employ in this experiment, the indicator used is K_2CrO_4. The chromate ion present in solutions of this substance will react with silver ion to form a red precipitate of Ag_2CrO_4. Under the conditions of the titration, the Ag^+ added to the solution reacts preferentially with Cl^- until that ion is essentially quantitatively removed from the system, at which point Ag_2CrO_4 begins to precipitate and the solution color changes from yellow to buff. The end point of the titration is that point at which the color change is first observed.

In this experiment, weighed samples containing an unknown percentage of chloride will be titrated with a standardized solution of $AgNO_3$, and the volumes of $AgNO_3$ solution required to reach the end point of each titration will be measured. Given the molarity of the $AgNO_3$

$$\text{no. of moles } Ag^+ = \text{no. of moles } AgNO_3 = M_{AgNO_3} \times V_{AgNO_3} \tag{2}$$

where the volume of $AgNO_3$ is expressed in liters and the molarity M_{AgNO_3} is in moles per liter of solution. At the end point of the titration,

$$\text{no. of moles } Ag^+ \text{ added} = \text{no. of moles } Cl^- \text{ present in unknown} \tag{3}$$

$$\text{no. of grams } Cl^- \text{ present} = \text{no. of moles } Cl^- \text{ present} \times MM \text{ Cl} \tag{4}$$

$$\% \text{ Cl} = \frac{\text{no. of grams } Cl^-}{\text{no. of grams unknown}} \times 100 \tag{5}$$

Experimental Procedure

Obtain from the stockroom a buret and a vial containing a sample of an unknown solid chloride. Weigh out accurately on the analytical balance three samples of the chloride, each sample weighing about 0.2 grams. This weighing is best done by accurately weighing the vial and its contents and pouring out the sample a little at a time into a 250-mL Erlenmeyer flask until the vial has lost about 0.2 g of chloride sample. Again weigh the sample vial accurately to obtain the exact amount of chloride sample poured into the flask. Put two other samples of similar mass into clean, dry, small beakers, weighing the vial accurately after the size of each sample has been decided upon. Add 50 mL of distilled water to the flask to dissolve the sample and add 6 drops of 0.5 M K_2CrO_4 indicator solution. Using the graduated cylinder at the reagent shelf, measure out about 100 mL of the standardized $AgNO_3$ solution into a clean *dry* 125-mL Erlenmeyer flask. This will be your total supply for the entire experiment so do not waste it. Clean your buret thoroughly with detergent solution and rinse it with distilled water. Pour three successive 2- or 3-mL portions of the $AgNO_3$ solution into the buret and tip it back and forth to rinse the inside walls. Allow the $AgNO_3$ solution to drain out the buret tip completely each time. Fill the buret with the $AgNO_3$ solution. Open the buret stopcock momentarily to flush any air bubbles out of the tip of the buret. Be sure your stopcock fits snugly and that the buret does not leak. (See Appendix IV for procedures regarding titrations with a buret.)

Read the initial buret level to 0.02 mL. You may find it useful when making readings to put a white card marked with a thick black stripe behind the meniscus. If the black line is held just below the level to be read, its reflection in the surface of the meniscus will help you obtain an accurate reading. Begin to add the $AgNO_3$ solution to the chloride solution in the Erlenmeyer flask. A white precipitate of AgCl will form immediately, and the amount will increase during the course of the titration. At the beginning of the titration, you can add the $AgNO_3$ fairly rapidly, a few milliliters at a time, swirling the flask as best you can to mix the solution. You will find that at the point where the $AgNO_3$ hits the solution, there will be a red spot of Ag_2CrO_4, which disappears when you stop adding nitrate and swirl the flask. As you proceed with the titration, the red spot will persist more and more, since the amount of excess chloride ion, which reacts with the Ag_2CrO_4 to form AgCl, will slowly decrease. Gradually decrease the rate at which you add $AgNO_3$ as the red color becomes stronger. At some stage you may find it convenient to set your buret stopcock to deliver $AgNO_3$ slowly, drop by drop, while you swirl the flask. When you are near the end point, add the $AgNO_3$ drop by drop, swirling between drops. The end point of the titration is that point where the mixture first takes on a permanent buff or reddish-yellow color that does not revert to pure yellow on swirling. If you are careful, you can hit the end point within 1 drop of $AgNO_3$. When you have reached the end point, stop the titration and record the buret level.

Pour the solution you have just titrated into another 125-mL Erlenmeyer flask or into a 250-mL beaker. To that solution add a few milliliters of 0.1 M NaCl; the color of the mixture should go back to the original yellow. Use the color of this mixture as a reference against which you compare your samples in the remaining titrations.

Rinse out the 250-mL Erlenmeyer flask in which you carried out the titration. Take your second sample and carefully pour it from the beaker into the Erlenmeyer flask. Wash out the beaker a few times with distilled water from your wash bottle and pour the washings into the flask. *All of the sample* must be transferred if the analysis is to be accurate. Add water to the flask to a volume of about 50 mL and swirl to dissolve the solid. Refill your buret, take a volume reading, add the indicator, and proceed to titrate to an end point as before. This titration should be more accurate than the first, since the volume of $AgNO_3$ used is proportional to sample size and therefore can be estimated rather well on the basis of the relative masses of the two samples. In addition, you have a reference for color comparison that should make it easier to recognize when a color change has occurred.

Titrate the third sample as you did the second. With care it should be possible to obtain volume-mass ratios that agree to within less than 1% in the last two titrations.

Optional Experiment: Find the mass percent Cl⁻ or NaCl in a bouillon cube, or canned chicken and rice or French onion soup. Compare your results with those on the label.

DISPOSAL OF REACTION PRODUCTS. Silver nitrate is very expensive. Pour all titrated solutions and any $AgNO_3$ remaining in your buret or flask into the waste bottles provided unless directed otherwise by your instructor.

Experiment 7

Data and Calculations: Analysis of an Unknown Chloride

Molarity of standard $AgNO_3$ solution _____ M

	I	II	III
Mass of vial and chloride unknown	_____ g	_____ g	_____ g
Mass of vial less sample	_____ g	_____ g	_____ g
Initial buret reading	_____ mL	_____ mL	_____ mL
Final buret reading	_____ mL	_____ mL	_____ mL
Mass of sample	_____ g	_____ g	_____ g
Volume of $AgNO_3$ used to titrate sample	_____ mL	_____ mL	_____ mL
No. of moles of $AgNO_3$ used to titrate sample	_____	_____	_____
No. of moles of Cl^- present in sample	_____	_____	_____
Mass of Cl^- present in sample	_____ g	_____ g	_____ g
Percentage of Cl^- in sample	_____ %	_____ %	_____ %

Mean value of percentage
of Cl^- in unknown _____ %

Unknown no. _____

Experiment 7

Advance Study Assignment: Analysis of an Unknown Chloride

1. A sample containing 0.213 g Cl^- is dissolved in 50.0 mL water.

    a. How many moles of Cl^- ion are in the solution?

                                                              _____ moles Cl^-

    b. What is the molarity of the Cl^- ion in the solution? ($M_{Cl^-} = n_{Cl^-} / V_{soln}$)

                                                              _____ M

2. A solid chloride sample weighing 0.2853 g required 43.75 mL of 0.05273 M $AgNO_3$ to reach the Ag_2CrO_4 end point.

    a. How many moles Cl^- ion were present in the sample? (Use Eqs. 2 and 3.)

                                                              _____ moles Cl^-

    b. How many grams Cl^- ion were present? (Use Eq. 4.)

                                                              _____ g Cl^-

    c. What was the mass percent Cl^- ion in the sample? (Use Eq. 5.)

                                                              _____ % Cl^-

3. How would the following errors affect the mass percent Cl^- obtained in Question 2c? Give your reasoning in each case.

    a. The student read the molarity of $AgNO_3$ as 0.05723 M instead of 0.05273 M.

    b. The student was past the end point of the titration when he took the final buret reading.

6

Molar Mass of a Volatile Liquid

One of the important applications of the Ideal Gas Law is found in the experimental determination of the molar masses of gases and vapors. In order to measure the molar mass of a gas or vapor we need simply to determine the mass of a given sample of the gas under known conditions of temperature and pressure. If the gas obeys the Ideal Gas Law,

$$PV = nRT \tag{1}$$

If the pressure P is in atmospheres, the volume V in liters, the temperature T in K, and the amount n in moles, then the gas constant R is equal to 0.0821 L atm/(mole K).

From the measured values of P, V, and T for a sample of gas we can use Equation 1 to find the number of moles of gas in the sample. The molar mass in grams, MM, is equal to the mass g of the gas sample divided by the number of moles n.

$$n = \frac{PV}{RT} \quad MM = \frac{g}{n} \tag{2}$$

This experiment involves measuring the molar mass of a volatile liquid by using Equation 2. A small amount of the liquid is introduced into a weighed flask. The flask is then placed in boiling water, where the liquid will vaporize completely, driving out the air and filling the flask with vapor at barometric pressure and the temperature of the boiling water. If we cool the flask so that the vapor condenses, we can measure the mass of the vapor and calculate a value for MM.

WEAR YOUR SAFETY GLASSES WHILE
PERFORMING THIS EXPERIMENT

Experimental Procedure*

Obtain a special round-bottomed flask, a stopper and cap, and an unknown liquid from the storeroom. Support the flask on an evaporating dish or in a beaker at all times. If you should break or crack the flask, report it to your instructor immediately so that it can be repaired. With the stopper loosely inserted in the neck of the flask, weigh the empty, dry flask on the analytical balance. Use a copper loop, if necessary, to suspend the flask from the hook supporting the balance pan. With an automatic balance, use a cork ring to support the flask.

Pour about half your unknown liquid, about 5 mL, into the flask. Assemble the apparatus as shown in Figure 9.1. Place the cap on the neck of the flask. Add a few boiling chips to the water in the 600-mL beaker and heat the water to the boiling point. Watch the liquid level in your flask; the level should gradually drop as vapor escapes through the cap. After all the liquid has disappeared and no more vapor comes out of the cap, continue to boil the water gently for 5 to 8 minutes. Measure the temperature of the boiling water. Shut off the burner and wait until the water has stopped boiling (about $\frac{1}{2}$ minute) and then loosen the clamp holding the flask in place. Slide out the flask, remove the cap, and *immediately* insert the stopper used previously.

Remove the flask from the beaker of water, holding it by the neck, which will be cooler. Immerse the flask in a beaker of cool water to a depth of about 5 cm. After holding the flask in the water for about 2 minutes to allow it to cool, carefully remove the stopper *for not more than a second or two* to allow air to enter, and again insert the stopper. (As the flask cools the vapor inside condenses and the pressure drops, which explains why air rushes in when the stopper is removed.)

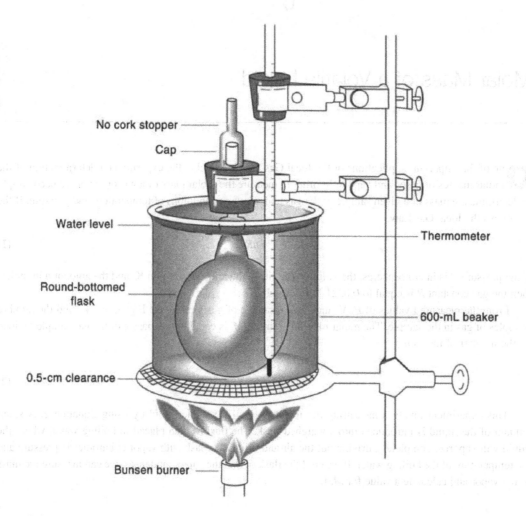

No cork stopper

Cap

Water level

Round-bottomed flask

0.5-cm clearance

Thermometer

600-mL beaker

Bunsen burner

Figure 9.1

Dry the flask with a towel to remove the surface water and let it cool to room temperature. Loosen the stopper momentarily to equalize any pressure differences, and reweigh the flask. Read the atmospheric pressure from the barometer.

Repeat the procedure using another 5 mL of your liquid sample.

You may obtain the volume of the flask from your instructor. Alternatively, he may direct you to measure its volume by weighing the flask stoppered and full of water on a top-loading balance. *Do not* fill the flask with water unless specifically told to do so.

When you have completed the experiment, return the flask to the storeroom; do not attempt to wash or clean it in any way.

Experiment 9

Data and Calculations: Molar Mass of a Volatile Liquid

	Trial 1	**Trial 2**
Unknown no.	_____	
Mass of flask and stopper	_____ g	_____ g
Mass of flask, stopper, and condensed vapor	_____ g	_____ g
Mass of flask, stopper, and water (see directions)	_____ g	_____ g
Temperature of boiling water bath	_____ °C	_____ °C
Barometric pressure	_____ mm Hg	_____ mm Hg

Calculations and Results

	Trial 1	**Trial 2**
Pressure of vapor, P	_____ atm	_____ atm
Volume of flask (volume of vapor), V	_____ L	_____ L
Temperature of vapor, T	_____ K	_____ K
Mass of vapor, g	_____ g	_____ g
Number of moles of vapor, n	_____ M	_____ M
Molar mass of unknown, as found by substitution into Equation 2	_____ g	_____ g

Experiment 9

Advance Study Assignment: Molar Mass of a Volatile Liquid

1. A student weighs an empty flask and stopper and finds the mass to be 54.868 g. She then adds about 5 mL of an unknown liquid and heats the flask in a boiling water bath at 100°C. After all the liquid is vaporized, she removes the flask from the bath, stoppers it, and lets it cool. After it is cool, she momentarily removes the stopper, then replaces it and weighs the flask and condensed vapor, obtaining a mass of 55.496 g. The volume of the flask is known to be 235.7 mL. The barometric pressure in the laboratory that day is 738 mm Hg.

    a. What was the pressure of the vapor in the flask in atm?

    $$P = \text{_____} \text{ atm}$$

    b. What was the temperature of the vapor in K? the volume of the flask in liters?

    $$T = \text{_____} \text{ K} \qquad V = \text{_____} \text{ L}$$

    c. What was the mass of vapor that was present in the flask?

    $$g = \text{_____} \text{ grams}$$

    d. How many moles of vapor are present?

    $$n = \text{_____} \text{ moles}$$

    e. What is the mass of one mole of vapor (Eq. 2)?

    $$MM = \text{_____} \text{ g/mole}$$

2. How would each of the following procedural errors affect the results to be expected in this experiment? Give your reasoning in each case.

    a. All of the liquid was not vaporized when the flask was removed from the water bath.

    b. The flask was not dried before the final weighing with the condensed vapor inside.

    c. The flask was left open to the atmosphere while it was being cooled, and the stopper was inserted just before the final weighing.

    d. The flask was removed from the bath before the vapor had reached the temperature of the boiling water. All the liquid had vaporized.

7

The Atomic Spectrum of Hydrogen

When atoms are excited, either in an electric discharge or with heat, they tend to give off light. The light is emitted only at certain wavelengths that are characteristic of the atoms in the sample. These wavelengths constitute what is called the atomic spectrum of the excited element and reveal much of the detailed information we have regarding the electronic structure of atoms.

Atomic spectra are interpreted in terms of quantum theory. According to this theory, atoms can exist only in certain states, each of which has an associated fixed amount of energy. When an atom changes its state, it must absorb or emit an amount of energy that is just equal to the difference between the energies of the initial and final states. This energy may be absorbed or emitted in the form of light. The emission spectrum of an atom is obtained when excited atoms fall from higher to lower energy levels. Since there are many such levels, the atomic spectra of most elements are very complex.

Light is absorbed or emitted by atoms in the form of photons, each of which has a specific amount of energy, ϵ. This energy is related to the wavelength of light by the equation

$$\epsilon_{\text{photon}} = \frac{hc}{\lambda} \tag{1}$$

where h is Planck's constant, 6.62608×10^{-34} joule seconds, c is the speed of light, 2.997925×10^8 meters per second, and λ is the wavelength, in meters. The energy ϵ_{photon} is in joules and is the energy given off by one atom when it jumps from a higher to a lower energy level. Since total energy is conserved, the change in energy of the atom, $\Delta\epsilon_{\text{atom}}$, must equal the energy of the photon emitted:

$$\Delta\epsilon_{\text{atom}} = \epsilon_{\text{photon}} \tag{2}$$

where $\Delta\epsilon_{\text{atom}}$ is equal to the energy in the upper level minus the energy in the lower one. Combining Equations 1 and 2, we obtain the relation between the change in energy of the atom and the wavelength of light associated with that change:

$$\Delta\epsilon_{\text{atom}} = \epsilon_{\text{upper}} - \epsilon_{\text{lower}} = \epsilon_{\text{photon}} = \frac{hc}{\lambda} \tag{3}$$

The amount of energy in a photon given off when an atom makes a transition from one level to another is very small, of the order of 1×10^{-19} joules. This is not surprising since, after all, atoms are very small particles. To avoid such small numbers, we will work with 1 mole of atoms, much as we do in dealing with energies involved in chemical reactions. To do this we need only to multiply Equation 3 by Avogadro's number, N. Let

$$N\Delta\epsilon = \Delta E = N\epsilon_{\text{upper}} - N\epsilon_{\text{lower}} = E_{\text{upper}} - E_{\text{lower}} = \frac{Nhc}{\lambda}$$

Substituting the values for N, h, and c, and expressing the wavelength in nanometers rather than meters (1 meter $= 1 \times 10^9$ nanometers), we obtain an equation relating energy change in kilojoules per mole of atoms to the wavelength of photons associated with such a change:

$$\Delta E = \frac{6.02214 \times 10^{23} \times 6.62608 \times 10^{-34}\,\text{J sec} \times 2.997925 \times 10^8\,\text{m/sec}}{\lambda\,(\text{in nm})} \times \frac{1 \times 10^9\,\text{nm}}{1\,\text{m}} \times \frac{1\,\text{kJ}}{1000\,\text{J}}$$

$$\Delta E = E_{\text{upper}} - E_{\text{lower}} = \frac{1.19627 \times 10^5\,\text{kJ/mole}}{\lambda\,(\text{in nm})} \quad \text{or} \quad \lambda\,(\text{in nm}) = \frac{1.19627 \times 10^5}{\Delta E\,(\text{in kJ/mole})} \tag{4}$$

Equation 4 is useful in the interpretation of atomic spectra. Say, for example, we study the atomic spectrum of sodium and find that the wavelength of the strong yellow line is 589.16 nm (see Fig. 11.1).

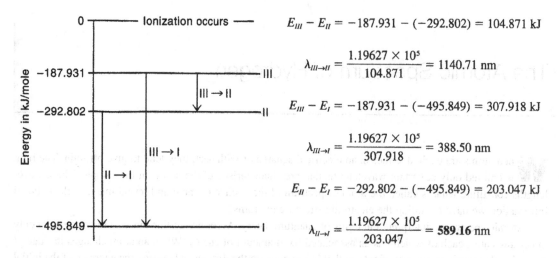

Figure 11.1 Calculation of wavelengths of spectral lines from energy levels of the sodium atom.

This line is known to result from a transition between two of the three lowest levels in the atom. The energies of these levels are shown in the figure. To make the determination of the levels which give rise to the 589.16 nm line, we note that there are three possible transitions, shown by downward arrows in the figure. We find the wavelengths associated with those transitions by first calculating ΔE ($E_{upper} - E_{lower}$) for each transition. Knowing ΔE we calculate λ by Equation 4. Clearly, the II $\to$ I transition is the source of the yellow line in the spectrum.

The simplest of all atomic spectra is that of the hydrogen atom. In 1886 Balmer showed that the lines in the spectrum of the hydrogen atom had wavelengths that could be expressed by a rather simple equation. Bohr, in 1913, explained the spectrum on a theoretical basis with his famous model of the hydrogen atom. According to Bohr's theory, the energies allowed to a hydrogen atom are given by the equation

$$\epsilon_n = \frac{-B}{n^2} \tag{5}$$

where B is a constant predicted by the theory and n is an integer, 1, 2, 3, . . . , called a quantum number. It has been found that all the lines in the atomic spectrum of hydrogen can be associated with energy levels in the atom which are predicted with great accuracy by Bohr's equation. When we write Equation 5 in terms of a mole of H atoms, and substitute the numerical value for B, we obtain

$$E_n = \frac{-1312.04}{n^2} \text{ kilojoules per mole, } n = 1, \ 2, \ 3, . . . \tag{6}$$

Using Equation 6 you can calculate, very accurately indeed, the energy levels for hydrogen. Transitions between these levels give rise to the wavelengths in the atomic spectrum of hydrogen. These wavelengths are also known very accurately. Given both the energy levels and the wavelengths, it is possible to determine the actual levels associated with each wavelength. In this experiment your task will be to make determinations of this type for the observed wavelengths in the hydrogen atomic spectrum that are listed in Table 11.1.

Table 11.1

Some Wavelengths (in nm) in the Spectrum of the Hydrogen Atom as Measured in a Vacuum					
Wavelength	**Assignment** $n_{hi} \to n_{lo}$	**Wavelength**	**Assignment** $n_{hi} \to n_{lo}$	**Wavelength**	**Assignment** $n_{hi} \to n_{lo}$
97.25	_____	410.29	_____	1005.2	_____
102.57	_____	434.17	_____	1094.1	_____
121.57	_____	486.27	_____	1282.2	_____
389.02	_____	656.47	_____	1875.6	_____
397.12	_____	954.86	_____	4052.3	_____

Experimental Procedure

There are several ways we might analyze an atomic spectrum, given the energy levels of the atom involved. A simple and effective method is to calculate the wavelengths of some of the lines arising from transitions between some of the lower energy levels, and see if they match those that are observed. We shall use this method in our experiment. All the data are good to at least five significant figures, so by using your hand calculator you should be able to make very accurate determinations. The Excel spreadsheet can be used to good advantage in this experiment. Your instructor may give you some suggestions on how you might proceed.

A. Calculations of the Energy Levels of the Hydrogen Atom

Given the expression for E_n in Equation 6, it is possible to calculate the energy for each of the allowed levels of the H atom starting with $n = 1$. Using your calculator, calculate the energy in kJ/mole of each of the 10 lowest levels of the H atom. Note that the energies are all negative, so that the *lowest* energy will have the *largest* allowed negative value. Enter these values in the table of energy levels, Table 11.2. On the energy level diagram provided, plot along the y axis each of the six lowest energies, drawing a horizontal line at the allowed level and writing the value of the energy alongside the line near the y axis. Write the quantum number associated with the level to the right of the line.

B. Calculation of the Wavelengths of the Lines in the Hydrogen Spectrum

The lines in the hydrogen spectrum all arise from jumps made by the atom from one energy level to another. The wavelengths in nm of these lines can be calculated by Equation 4, where ΔE is the difference in energy in kJ/mole between any two allowed levels. For example, to find the wavelength of the spectral line associated with a transition from the $n = 2$ level to the $n = 1$ level, calculate the difference, ΔE, between the energies of those two levels. Then substitute ΔE into Equation 4 to obtain this wavelength in nanometers.

Using the procedure we have outlined, calculate the wavelengths in nm of all the lines we have indicated in Table 11.3. That is, calculate the wavelengths of all the lines that can arise from transitions between any two of the six lowest levels of the H atom. Enter these values in Table 11.3.

C. Assignment of Observed Lines in the Hydrogen Spectrum

Compare the wavelengths you have calculated with those listed in Table 11.1. If you have made your calculations properly, your wavelengths should match, within the error of your calculation, several of those that are observed. On the line opposite each wavelength in Table 11.1, write the quantum numbers of the upper and lower states for each line whose origin you can recognize by comparison of your calculated values with the observed values. On the energy level diagram, draw a vertical arrow pointing down (light is emitted, $\Delta E < 0$) between those pairs of levels that you associate with any of the observed wavelengths. By each arrow write the wavelength of the line originating from that transition.

There are a few wavelengths in Table 11.1 that have not yet been calculated. Enter those wavelengths in Table 11.4. By assignments already made and by an examination of the transitions you have marked on the diagram, deduce the quantum states that are likely to be associated with the as yet unassigned lines. This is perhaps most easily done by first calculating the value of ΔE, which is associated with a given wavelength. Then find two values of E_n whose difference is equal to ΔE. The quantum numbers for the two E_n states whose energy difference is ΔE will be the ones that are to be assigned to the given wavelength. When you have found n_{hi} and n_{lo} for a wavelength, write them in Table 11.1 and Table 11.4; continue until all the lines in the table have been assigned.

D. The Balmer Series

This is the most famous series in the atomic spectrum of hydrogen. The lines in this series are the only ones in the spectrum that occur in the visible region. Your instructor may have a hydrogen source tube and a spectroscope with which you may be able to observe some of the lines in the Balmer series. In the Data and Calculations section are some questions you should answer relating to this series.

Experiment 11

Data and Calculations: The Atomic Spectrum of Hydrogen

A. Calculation of the Energy Levels of the Hydrogen Atom

Energies are to be calculated from Equation 6 for the 10 lowest energy states.

Table 11.2

Quantum Number, n	Energy, E_n, in kJ/mole	Quantum Number, n	Energy, E_n, in kJ/mole
____	____	____	____
____	____	____	____
____	____	____	____
____	____	____	____
____	____	____	____

B. Calculation of Wavelengths in the Spectrum of the H Atom

In the upper half of each box write ΔE, the difference in energy in kJ/mole between $E_{n_{hi}}$ and $E_{n_{lo}}$. In the lower half of the box, write λ in nm associated with that value of ΔE.

Table 11.3

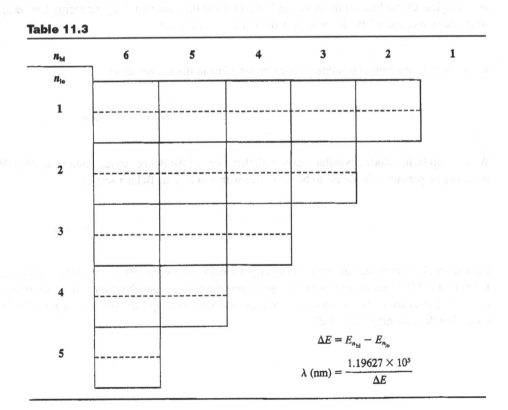

$$\Delta E = E_{n_{hi}} - E_{n_{lo}}$$

$$\lambda \text{ (nm)} = \frac{1.19627 \times 10^5}{\Delta E}$$

C. Assignment of Wavelengths

1. As directed in the procedure, assign n_{hi} and n_{lo} for each wavelength in Table 11.1 which corresponds to a wavelength calculated in Table 11.3.

2. List below any wavelengths you cannot yet assign.

Table 11.4

Wavelength λ Observed	ΔE Transition	Probable Transition $n_{hi} \to n_{lo}$	λ Calculated in nm (Eq. 4)
————	————	————	————
————	————	————	————
————	————	————	————
————	————	————	————

D. The Balmer Series

1. When Balmer found his famous series for hydrogen in 1886, he was limited experimentally to wavelengths in the visible and near ultraviolet regions from 250 nm to 700 nm, so all the lines in his series lie in that region. On the basis of the entries in Table 11.3 and the transitions on your energy level diagram, what common characteristic do the lines in the Balmer series have?

    What would be the longest possible wavelength for a line in the Balmer series?

    $$\lambda = \underline{\hspace{2cm}} \text{ nm}$$

    What would be the shortest possible wavelength that a line in the Balmer series could have? Hint: What is the largest possible value of ΔE to be associated with a line in the Balmer series?

    $$\lambda = \underline{\hspace{2cm}} \text{ nm}$$

    Fundamentally, why would any line in the hydrogen spectrum between 250 nm and 700 nm belong to the Balmer series? Hint: On the energy level diagram note the range of possible values of ΔE for transitions to the $n = 1$ level and to the $n = 3$ level. Could a spectral line involving a transition to the $n = 1$ level have a wavelength in the range indicated?

The Ionization Energy of Hydrogen

1. In the normal hydrogen atom the electron is in its lowest energy state, which is called the ground state of the atom. The maximum electronic energy that a hydrogen atom can have is 0 kJ/mole, at which point the electron would essentially be removed from the atom and it would become a H^+ ion. How much energy in kilojoules per mole does it take to ionize an H atom?

_____ kJ/mole

The ionization energy of hydrogen is often expressed in units other than kJ/mole. What would it be in joules per atom? in electron volts per atom? (1 ev = 1.602×10^{-19} J)

_____ J/atom; _____ ev/atom

Experiment 11

The Atomic Spectrum of Hydrogen: Energy Level Diagram

(Data and Calculations)

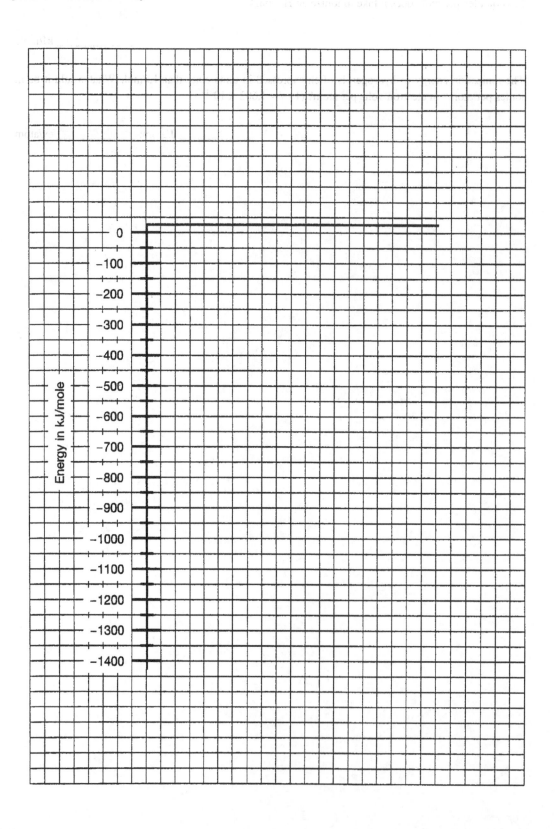

Experiment 11

Advance Study Assignment: The Atomic Spectrum of Hydrogen

1. The helium ion, He^+, has energy levels similar to those of the hydrogen atom, since both species have only one electron. The energy levels of the He^+ ion are given by the equation

$$E_n = -\frac{5248.16}{n^2}\,\text{kJ/mole} \quad n = 1,\ 2,\ 3,\ldots$$

a. Calculate the energies in kJ/mole for the four lowest energy levels of the He^+ ion.

$E_1 =$ _____ kJ/mole

$E_2 =$ _____ kJ/mole

$E_3 =$ _____ kJ/mole

$E_4 =$ _____ kJ/mole

b. One of the most important transitions for the He^+ ion involves a jump from the $n = 2$ to the $n = 1$ level. ΔE for this transition equals $E_2 - E_1$, where these two energies are obtained as in Part a. Find the value of ΔE in kJ/mole. Find the wavelength in nm of the line emitted when this transition occurs; use Equation 4 to make the calculation.

$\Delta E =$ _____ kJ/mole; $\lambda =$ _____ nm

c. Three of the strongest lines in the He^+ ion spectrum are observed at the following wavelengths: (1) 121.57 nm; (2) 164.12 nm; (3) 468.90 nm. Find the quantum numbers of the initial and final states for the transitions that give rise to these three lines. Do this by calculating, using Equation 4, the wavelengths of lines that can originate from transitions involving any two of the four lowest levels. You calculated one such wavelength in Part b. Make similar calculations with the other possible pairs of levels. When a calculated wavelength matches an observed one, write down n_{hi} and n_{lo} for that line. Continue until you have assigned all three of the lines.

(1) _____ → _____   (2) _____ → _____   (3) _____ → _____

Experiment 7.1

Advance Study Assignment: The Atomic Spectrum of Hydrogen

8

The Alkaline Earths and the Halogens—Two Families in the Periodic Table

The Periodic Table arranges the elements in order of increasing atomic number in horizontal rows of such length that elements with similar properties recur periodically; that is, they fall directly beneath each other in the Table. The elements in a given vertical column are referred to as a family or group. The physical and chemical properties of the elements in a given family change gradually as one goes from one element in the column to the next. By observing the trends in properties the elements can be arranged in the order in which they appear in the Periodic Table. In this experiment we will study the properties of the elements in two families in the Periodic Table, the alkaline earths (Group 2) and the halogens (Group 17).

The alkaline earths are all moderately reactive metals and include barium, beryllium, calcium, magnesium, radium, and strontium. (Since beryllium compounds are rarely encountered and often very poisonous, and radium compounds are highly radioactive, we will not include these two elements in this experiment.) All the alkaline earths exist in their compounds and in solution as M^{2+} cations (Mg^{2+}, Ca^{2+}, etc.). If a solution containing one of these cations is mixed with one containing an anion (CO_3^{2-}, SO_4^{2-}, IO_3^-, etc.), an alkaline earth salt will precipitate if the compound containing those two ions is insoluble.

For example:

$$M^{2+}(aq) + SO_4^{2-}(aq) \rightarrow MSO_4(s) \quad \text{if } MSO_4 \text{ is insoluble} \tag{1a}$$

$$M^{2+}(aq) + 2\ IO_3^-(aq) \rightarrow M(IO_3)_2(s) \quad \text{if } M(IO_3)_2 \text{ is insoluble} \tag{1b}$$

We would expect, and indeed observe, that the solubilities of the salt of the alkaline earth cations with any one of the given anions show a smooth trend consistent with the order of the cations in the Periodic Table. That is, as we go from one end of the alkaline earth family to the other, the solubilities of, say, the sulfate salts either gradually increase or decrease. Similar trends exist for the carbonates, oxalates, and iodates formed by those cations. By determining such trends in this experiment, you will be able to confirm the order of the alkaline earths in the Periodic Table.

The elementary halogens are also relatively reactive. They include astatine, bromine, chlorine, fluorine, and iodine. We will not study astatine and fluorine in this experiment, since the former is radioactive and the latter is too reactive to be safe. Unlike the alkaline earths, the halogen atoms tend to gain electrons, forming X^- anions (Cl^-, Br^-, etc.). Because of this property, the halogens are oxidizing agents, species that tend to oxidize (remove electrons from) other species. An interesting and simple example of the sort of reaction that may occur arises when a solution containing a halogen (Cl_2, Br_2, I_2) is mixed with a solution containing a halide ion (Cl^-, Br^-, I^-). Taking X_2 to be the halogen, and Y^- to be a halide ion, the following reaction may occur, in which another halogen, Y_2, is formed:

$$X_2(aq) + 2\ Y^-(aq) \rightarrow 2\ X^-(aq) + Y_2(aq) \tag{2}$$

The reaction will occur if X_2 is a better oxidizing agent than Y_2, since then X_2 can produce Y_2 by removing electrons from the Y^- ions. If Y_2 is a better oxidizing agent than X_2, Reaction 2 will not proceed but will be spontaneous in the opposite direction.

In this experiment we will mix solutions of halogens and halide ions to determine the relative oxidizing strengths of the halogens. These strengths show a smooth variation as one goes from one halogen to the next in the Periodic Table. We will be able to tell if a reaction occurs by the colors we observe. In water, and particularly in some organic solvents, the halogens have characteristic colors. The halide ions are colorless in water solution and insoluble in organic solvents. Bromine (Br_2) in hexane, C_6H_{14} (HEX), is orange, while Cl_2 and I_2 in that solvent have quite different colors.

Say, for example, we shake a water solution of Br_2 with a little hexane, which is lighter than and insoluble in water. The Br_2 is much more soluble in HEX than in water and goes into the HEX layer, giving it an orange color. To that mixture we add a solution containing a halide ion, say Cl^- ion, and mix well. If Br_2 is a better oxidizing agent than Cl_2, it will take electrons from the chloride ions and will be converted to bromide, Br^-, ions; the reaction would be

$$Br_2(aq) + 2\ Cl^-(aq) \rightarrow 2\ Br^-(aq) + Cl_2(aq) \qquad (3)$$

If the reaction occurs, the color of the HEX layer will of necessity change, since Br_2 will be used up and Cl_2 will form. The color of the HEX layer will go from orange to that of a solution of Cl_2 in HEX. If the reaction does *not* occur, the color of the HEX layer will remain orange. By using this line of reasoning, and by working with the possible mixtures of halogens and halide ions, you should be able to arrange the halogens in order of increasing oxidizing power, which must correspond to their order in the Periodic Table.

One difficulty that you may have in this experiment involves terminology rather than actual chemistry. You must learn to distinguish the halogen *elements* from the halide *ions*, since the two kinds of species are not at all the same, even though their names are similar:

Elementary halogens	Halide ions
Bromine, Br_2	Bromide ion, Br^-
Chlorine, Cl_2	Chloride ion, Cl^-
Iodine, I_2	Iodide ion, I^-

The *halogens* are molecular substances and oxidizing agents, and all have odors. They are only slightly soluble in water and are much more soluble in HEX, where they have distinct colors. The *halide ions* exist in solution only in water, have no color or odor, and are *not* oxidizing agents. They do not dissolve in HEX.

Given the solubility properties of the alkaline earth cations, and the oxidizing power of the halogens, it is possible to develop a systematic procedure for determining the presence of any Group 2 cation and any Group 17 anion in a solution. In the last part of this experiment you will be asked to set up such a procedure and use it to establish the identity of an unknown solution containing a single alkaline earth halide.

WEAR YOUR SAFETY GLASSES WHILE PERFORMING THIS EXPERIMENT

Experimental Procedure

I. Relative Solubilities of Some Salts of the Alkaline Earths

To each of four small test tubes add about 1 mL (approximately 12 drops) of 1 M H_2SO_4. Then add 1 mL of 0.1 M solutions of the nitrate salts of barium, calcium, magnesium, and strontium to those tubes, one solution to a tube. Stir each mixture with your glass stirring rod, rinsing the rod in a beaker of distilled water between stirs. Record your results on the solubilities of the sulfates of the alkaline earths in the Table, noting whether a precipitate forms, and any characteristics (such as color, amount, size of particles, and settling tendencies) that might distinguish it.

Rinse out the test tubes, and to each add 1 mL 1 M Na_2CO_3. Then add 1 mL of the solutions of the alkaline earth salts, one solution to a tube, as before. Record your observations on the solubility properties of the carbonates of the alkaline earth cations. Rinse out the tubes, and test for the solubilities of the oxalates of these cations, using 0.25 M $(NH_4)_2C_2O_4$ as the precipitating reagent. Finally, determine the relative solubilities of the iodates of the alkaline earths, using 1 mL 0.1 M KIO_3 as the test reagent.

II. Relative Oxidizing Powers of the Halogens

In a small test tube place a few milliliters of bromine-saturated water and add 1 mL of hexane. Stopper the test tube and shake until the bromine color is mostly in the HEX layer. **CAUTION:** *Avoid breathing the halogen vapors. Don't use your finger to stopper the tube, since a halogen solution can give you a bad chemical burn.* Repeat the experiment using chlorine water and iodine water with separate samples of HEX, noting any color changes as the bromine, chlorine, and iodine are extracted from the water layer into the HEX layer.

To each of three small test tubes add 1 mL bromine water and 1 mL HEX. Then add 1 mL 0.1 M NaCl to the first test tube, 1 mL 0.1 M NaBr to the second, and 1 mL 0.1 M NaI to the third. Stopper each tube and shake it. Note the color of the HEX phase above each solution. If the color is not that of Br_2 in HEX, a reaction indeed occurred, and Br_2 oxidized that anion, producing the halogen. In such a case, Br_2 is a stronger oxidizing agent than the halogen that was produced.

Rinse out the tubes, and this time add 1 mL chlorine water and 1 mL HEX to each tube. Then add 1 mL of the 0.1 M solutions of the sodium halide salts, one solution to a tube, as before. Stopper each tube and shake, noting the color of the HEX layer after shaking. Depending on whether the color is that of Cl_2 in HEX or not, decide whether Cl_2 is a better oxidizing agent than Br_2 or I_2. Again, rinse out the tubes, and add 1 mL iodine water and 1 mL HEX to each. Test each tube with 1 mL of a sodium halide salt solution, and determine whether I_2 is able to oxidize Cl^- or Br^- ions. Record all your observations in the Table.

III. Identification of an Alkaline Earth Halide

Your observations on the solubility properties of the alkaline earth cations should allow you to develop a method for determining which of those cations is present in a solution containing one Group 2 cation and no other cations. The method will involve testing samples of the solution with one or more of the reagents you used in Part I. Indicate on the Data page how you would proceed.

In a similar way you can determine which halide ion is present in a solution containing only one such anion and no others. There you will need to test a solution of an oxidizing halogen with your unknown to see how the halide ion is affected. From the behavior of the halogen-halide ion mixtures you studied in Part II you should be able to identify easily the particular halide that is present. Describe your method on the Data page, obtain an unknown solution of an alkaline earth halide, and then use your procedure to determine the cation and anion that it contains.

IV. Microscale Procedure for Determining Solubilities of Alkaline Earth Salts Optional

Your instructor may have you carry out Part I of this experiment by a microscale approach. This method uses much smaller amounts of reagents. Plastic well plates are employed as containers, and reagents are measured out with small Beral pipettes.

Using Beral pipettes, add four drops 0.1 M $Ba(NO_3)_2$, barium nitrate, to wells A1–A4, four drops to each well. Similarly, add four drops 0.1 M $Ca(NO_3)_2$, calcium nitrate, to wells B1–B4; four drops of 0.1 M $Mg(NO_3)_2$, magnesium nitrate, to wells C1–C4, and four drops 0.1 M $Sr(NO_3)_2$, strontium nitrate, to wells D1–D4.

Then, with another Beral pipette, add four drops 1 M H_2SO_4, sulfuric acid, to wells A1–D1. In the Table, record your results on the solubilities of the sulfates of the alkaline earths. Note whether a precipitate formed, and any characteristics, such as amount, size of particles, and cloudiness, which might distinguish it.

With a different Beral pipette, add four drops of 1 M Na_2CO_3, sodium carbonate, to wells A2–D2. Record your observations on the solubilities of the carbonates of the alkaline earths. Then carry out the same sort of tests with 0.25 M $(NH_4)_2C_2O_4$, ammonium oxalate, in wells A3–D3, and finally with 0.1 M KIO_3, potassium iodate, in wells A4–D4. Note all of your observations in the Table.

Optional Experiment: Dissolve a sample of limestone in acid, and determine whether it contains Mg^{2+} as well as Ca^{2+} ions. Estimate the relative concentrations of the two ions.

DISPOSAL OF REACTION PRODUCTS: Dispose of the reaction products from this experiment as directed by your instructor.

Experiment 12

Observations and Conclusions: The Alkaline Earths and the Halogens

I. Solubilities of Salts of the Alkaline Earths

	1 M H_2SO_4	1 M Na_2CO_3	0.25 M $(NH_4)_2C_2O_4$	0.1 M KIO_3
$Ba(NO_3)_2$				
$Ca(NO_3)_2$				
$Mg(NO_3)_2$				
$Sr(NO_3)_2$				

Key: P = precipitate forms; S = no precipitate forms.

Note any distinguishing characteristics of precipitate, such as amount and degree of cloudiness.

Consider the relative solubilities of the Group 2 cations in the various precipitating reagents. On the basis of the trends you observed, list the four alkaline earths in the order in which they should appear in the Periodic Table. *Start with the one which forms the most soluble oxalate.*

most soluble _____ _____ _____ _____ least soluble

Why did you arrange the elements as you did? Is the order consistent with the properties of the cations in all of the participating reagents?

II. Relative Oxidizing Powers of the Halogens

a. Color of the halogen in solution:

	Br_2	Cl_2	I_2
HEX			
Water			

b. Reactions between halogens and halides:

	Br$^-$	Cl$^-$	I$^-$
Br$_2$			
Cl$_2$			
I$_2$			

State your observations with each mixture, noting the initial and final colors of the HEX layer, and which halogen ends up in the HEX layer. Key: R = reaction occurs; NR = no reaction occurs.

Rank the halogens in order of their increasing oxidizing power.

weakest _____ _____ _____ strongest

Is this their order in the Periodic Table?

III. Identification of an Alkaline Earth Halide

Procedure for identifying the Group 2 cation:

Procedure for identifying the Group 17 anion:

Observations on unknown alkaline earth halide solution:

Cation present_____

Anion present_____

Unknown no._____

Experiment 12

Advance Study Assignment: The Alkaline Earths and the Halogens

1. All of the common noble gases are monatomic and low-boiling. Their boiling points in °C are: Ne, −245; Ar, −186; Kr, −152; Xe, −107. Using the Periodic Table, predict as best you can the molecular formula and boiling point of radon, Rn, the only radioactive element in this family.

_____ _____ °C

2. Substances A, B, and C can all act as oxidizing agents. In solution, A is green, B is yellow, and C is red. In the reactions in which they participate, they are reduced to A⁻, B⁻, and C⁻ ions, all of which are colorless. When a solution of C is mixed with one containing A⁻ ions, the color changes from red to green.

Which species is oxidized? _____

Which is reduced? _____

When a solution of C is mixed with one containing B⁻ ions, the color remains red.

Is C a better oxidizing agent than A? _____

Is C a better oxidizing agent than B? _____

Arrange A, B, and C in order of increasing strengths as oxidizing agents.

3. You are given an unknown, colorless, solution that may contain only one salt from the following set: NaA, NaB, NaC. In solution each salt dissociates completely into the Na⁺ ion and the anion A⁻, B⁻, or C⁻, whose properties are given in Problem 2. The Na⁺ ion is effectively inert. Given the availability of solutions of A, B, and C, develop a simple procedure for identifying the salt that is present in your unknown.

9

The Geometrical Structure of Molecules— An Experiment Using Molecular Models

Many years ago it was observed that in many of its compounds the carbon atom formed four chemical linkages to other atoms. As early as 1870, graphic formulas of carbon compounds were drawn as shown:

$$
\begin{array}{ccc}
& H & \\
& | & \\
H-&C&-H \\
& | & \\
& H &
\end{array}
\qquad
\begin{array}{c}
H \quad H \\
| \quad\; | \\
C = C \\
| \quad\; | \\
H \quad H
\end{array}
$$

methane   ethylene

Although such drawings as these would imply that the atom-atom linkages, indicated by valence strokes, lie in a plane, chemical evidence, particularly the existence of only one substance with the graphic formula

$$
\begin{array}{c}
Cl \\
| \\
H-C-Cl \\
| \\
H
\end{array}
$$

requires that the linkages be directed toward the corners of a tetrahedron, at the center of which is the carbon atom.

The physical significance of the chemical linkages between atoms, expressed by the lines or valence strokes in molecular structure diagrams, became evident soon after the discovery of the electron. In 1916 in a classic paper, G. N. Lewis suggested, on the basis of chemical evidence, that the single bonds in graphic formulas involve two electrons and that an atom tends to hold eight electrons in its outermost or valence shell.

Lewis' proposal that atoms generally have eight electrons in their outer shells proved to be extremely useful and has come to be known as the octet rule. It can be applied to many atoms, but is particularly important in the treatment of covalent compounds of atoms in the second row of the Periodic Table. For atoms such as carbon, oxygen, nitrogen, and fluorine, the eight valence electrons occur in pairs that occupy tetrahedral positions around the central atom core. Some of the electron pairs do not participate directly in chemical bonding and are called unshared or nonbonding pairs; however, the structures of compounds containing such unshared pairs reflect the tetrahedral arrangement of the four pairs of valence shell electrons. In the H_2O molecule, which obeys the octet rule, the four pairs of electrons around the central oxygen atom occupy essentially tetrahedral positions; there are two unshared nonbonding pairs and two bonding pairs that are shared by the O atom and the two H atoms. The H—O—H bond angle is nearly but not exactly tetrahedral since the properties of shared and unshared pairs of electrons are not exactly alike.

$$
\begin{array}{c}
\cdot \ddot{O} \cdot \\
\diagup \quad \diagdown \\
H \qquad H
\end{array}
$$

Most molecules obey the octet rule. Essentially, all organic molecules obey the rule, and so do most inorganic molecules and ions. For species that obey the octet rule it is possible to draw electron-dot, or Lewis, structures. The previous drawing of the H_2O molecule is an example of a Lewis structure. Here are several others:

$$
\begin{array}{ccccc}
:\ddot{Cl}: & & & H \quad H & \\
| & & & | \quad\; | & \\
H-C-H & H-\ddot{N}-H & :\ddot{O}-H^{-} & H-C-C-H & H-C=C-H \\
| & | & & | \quad\; | & | \qquad | \\
:\ddot{Cl}: & H & & H \quad H & H \qquad H
\end{array}
$$

In each of the previous structures there are eight electrons around each atom (except for H atoms, which always have two electrons). There are two electrons in each bond. When counting electrons in these structures, one considers the electrons in a bond between two atoms as belonging to the atom under consideration. In the CH_2Cl_2 molecule just shown, for example, the Cl atoms each have eight electrons, including the two in the single bond to the C atom. The C atom also has eight electrons, two from each of the four bonds to that atom. The bonding and nonbonding electrons in Lewis structures are all from the *outermost* shells of the atoms involved, and are the so-called valence electrons of those atoms. For the main group elements, the number of valence electrons in an atom is equal to the last digit in the group number of the element in the Periodic Table. Carbon, in Group 4, has four valence electrons in its atoms; hydrogen, in Group 1, has one; chlorine, in Group 17, has seven valence electrons. In an octet rule structure the valence electrons from all the atoms are arranged in such a way that each atom, except hydrogen, has eight electrons.

Often it is quite easy to construct an octet rule structure for a molecule. Given that an oxygen atom has six valence electrons (Group 6) and a hydrogen atom has one, it is clear that one O and two H atoms have a total of eight valence electrons; the octet rule structure for H_2O, which we discussed earlier, follows by inspection. Structures like that of H_2O, involving only single bonds and nonbonding electron pairs, are common. Sometimes, however, there is a "shortage" of electrons; that is, it is not possible to construct an octet rule structure in which all the electron pairs are either in single bonds or are nonbonding. C_2H_4 is a typical example of such a species. In such cases, octet rule structures can often be made in which two atoms are bonded by two pairs, rather than one pair, of electrons. The two pairs of electrons form a double bond. In the C_2H_4 molecule, shown above, the C atoms each get four of their electrons from the double bond. The assumption that electrons behave this way is supported by the fact that the $C=C$ double bond is both shorter and stronger than the $C-C$ single bond in the C_2H_6 molecule (see previous example). Double bonds, and triple bonds, occur in many molecules, usually between C, O, N, and/or S atoms.

Lewis structures can be used to predict molecular and ionic geometries. All that is needed is to assume that the four pairs of electrons around each atom are arranged tetrahedrally. We have seen how that assumption leads to the correct geometry for H_2O. Applying the same principle to the species whose Lewis structures we listed earlier, we would predict, correctly, that the CH_2Cl_2 molecule would be tetrahedral (roughly anyway), that NH_3 would be pyramidal (with the nonbonding electron pair sticking up from the pyramid made from the atoms), that the bond angles in C_2H_6 are all tetrahedral, and that the C_2H_4 molecule is planar (the two bonding pairs in the double bond are in a sort of banana bonding arrangement above and below the plane of the molecule). In describing molecular geometry we indicate the positions of the atomic nuclei, not the electrons. The NH_3 molecule is pyramidal, not tetrahedral.

It is also possible to predict polarity from Lewis structures. Polar molecules have their center of positive charge at a different point than their center of negative charge. This separation of charges produces a dipole moment in the molecule. Covalent bonds between different kinds of atoms are polar; all heteronuclear diatomic molecules are polar. In some molecules the polarity from one bond may be canceled by that from others. Carbon dioxide, CO_2, which is linear, is a nonpolar molecule. Methane, CH_4, which is tetrahedral, is also nonpolar. Among the species whose Lewis structures we have listed, we find that H_2O, CH_2Cl_2, NH_3, and OH^- are polar. C_2H_6 and C_2H_4 are nonpolar.

For some molecules with a given molecular formula, it is possible to satisfy the octet rule with different atomic arrangements. A simple example would be

$$
\begin{array}{cccc}
\text{H} & \text{H} & & \text{H} & \text{H} \\
| & | & & | & | \\
\text{H}-\text{C}-\text{C}-\ddot{\text{O}}-\text{H} & \text{and} & \text{H}-\text{C}-\ddot{\text{O}}-\text{C}-\text{H} \\
| & | & & | & | \\
\text{H} & \text{H} & & \text{H} & \text{H}
\end{array}
$$

The two molecules are called isomers of each other, and the phenomenon is called isomerism. Although the molecular formulas of both substances are the same, C_2H_6O, their properties differ markedly because of their different atomic arrangements.

Isomerism is very common, particularly in organic chemistry, and when double bonds are present, isomerism can occur in very small molecules:

$$
\begin{array}{ccccc}
\overset{H}{\underset{\ddot{Cl}}{\diagdown}}C{=}C\overset{H}{\underset{\ddot{Cl}}{\diagup}} & \text{and} & \overset{\ddot{Cl}}{\underset{H}{\diagdown}}C{=}C\overset{H}{\underset{\ddot{Cl}}{\diagup}} & \text{and} & \overset{\ddot{Cl}}{\underset{\ddot{Cl}}{\diagdown}}C{=}C\overset{H}{\underset{H}{\diagup}}
\end{array}
$$

The first two isomers result from the fact that there is no rotation around a double bond, although such rotation can occur around single bonds. The third isomeric structure cannot be converted to either of the first two without breaking bonds.

With certain molecules, given a fixed atomic geometry, it is possible to satisfy the octet rule with more than one bonding arrangement. The classic example is benzene, whose molecular formula is C_6H_6:

[Two resonance structures of benzene drawn with alternating single and double bonds, labeled "and" between them.]

These two structures are called resonance structures, and molecules such as benzene, which have two or more resonance structures, are said to exhibit resonance. The actual bonding in such molecules is thought to be an average of the bonding present in the resonance structures. The stability of molecules exhibiting resonance is found to be higher than that anticipated for any single resonance structure.

Although the conclusions we have drawn regarding molecular geometry and polarity can be obtained from Lewis structures, it is much easier to draw such conclusions from models of molecules and ions. The rules we have cited for octet rule structures transfer readily to models. In many ways the models are easier to construct than are the drawings of Lewis structures on paper. In addition, the models are three-dimensional and hence much more representative of the actual species. Using the models, it is relatively easy to see both geometry and polarity, as well as to deduce Lewis structures. In this experiment you will assemble models for a sizeable number of common chemical species and interpret them in the ways we have discussed.

Experimental Procedure

IN THIS EXPERIMENT YOUR INSTRUCTOR MAY ALLOW
YOU TO WORK WITHOUT SAFETY GLASSES

In this experiment you may work in pairs during the first portion of the laboratory period.

The models you will use consist of drilled wooden balls, short sticks, and springs. The balls represent atomic nuclei surrounded by the inner electron shells. The sticks and springs represent electron pairs and fit in the holes in the wooden balls. The model (molecule or ion) consists of wooden balls (atoms) connected by sticks or springs (chemical bonds). Some sticks may be connected to only one atom (nonbonding pairs).

In this experiment we will deal with atoms that obey the octet rule; such atoms have four electron pairs around the central core and will be represented by balls with four tetrahedral holes in which there are four sticks or springs. The only exception will be hydrogen atoms, which share two electrons in covalent compounds, and which will be represented by balls with a single hole in which there is a single stick.

In assembling a molecular model of the kind we are considering, it is possible, indeed desirable, to proceed in a systematic manner. We will illustrate the recommended procedure by developing a model for a molecule with the formula CH_2O.

1. Determine the total number of valence electrons in the species. This is easily done once you realize that the number of valence electrons on an atom is equal to the number of the group to which the atom belongs in the Periodic Table. For CH_2O,

C   Group 4   H Group 1   O   Group 6

   Therefore each carbon atom in a molecule or ion contributes four electrons, each hydrogen atom one electron, and each oxygen atom six electrons. The total number of valence electrons equals the sum of the valence electrons on all of the atoms in the species being studied. For CH_2O this total would be $4 + (2 \times 1) + 6$, or 12 valence electrons. If we are working with an ion, we add one electron for each negative charge or subtract one for each positive charge on the ion.

2. Select wooden balls and sticks to represent the atoms and electron pairs in the molecule. You should use four-holed balls for the carbon atom and the oxygen atom, and one-holed balls to represent the hydrogen atoms. Since there are 12 valence electrons in the molecule and electrons occur in pairs, you will need six sticks to represent the six electron pairs. The sticks will serve both as bonds between atoms and as nonbonding electron pairs.

3. Connect the balls with some of the sticks. (Assemble a skeleton structure for the molecule, joining atoms by single bonds.) In some cases this can only be done in one way. Usually, however, there are various possibilities, some of which are more reasonable than others. In CH_2O the model can be assembled by connecting the two H atom balls to the C atom ball with two of the available sticks, and then using a third stick to connect the C atom and O atom balls.

4. The next step is to use the sticks that are left over in such a way as to fill all the remaining holes in the balls. (Distribute the electron pairs so as to give each atom eight electrons and so satisfy the octet rule.) In the model we have assembled, there is one unfilled hole in the C atom ball, three unfilled holes in the O atom ball, and three available sticks. An obvious way to meet the required condition is to use two sticks to fill two of the holes in the O atom ball, and then use two springs instead of two sticks to connect the C atom and O atom balls. The model as completed is shown in Figure 13.1.

5. Interpret the model in terms of the atoms and bonds represented. The sticks and spatial arrangement of the balls will closely correspond to the electronic and atomic arrangement in the molecule. Given our model, we would describe the CH_2O molecule as being planar with single bonds between carbon and hydrogen atoms and a double bond between the C and O atoms. The H—C—H angle is approximately tetrahedral. There are two nonbonding electron pairs on the O atom. Since all bonds are polar and the molecular symmetry

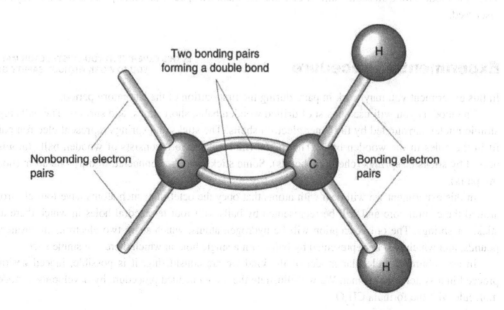

does not cancel the polarity in CH_2O, the molecule is polar. The Lewis structure of the molecule is given below:

(The compound having molecules with the formula CH_2O is well known and is called formaldehyde. The bonding and structure in CH_2O are as given by the model.)

6. Investigate the possibility of the existence of isomers or resonance structures. It turns out that in the case of CH_2O one can easily construct an isomeric form that obeys the octet rule, in which the central atom is oxygen rather than carbon. It is found that this isomeric form of CH_2O does not exist in nature. As a general rule, carbon atoms almost always form a total of four bonds; put another way, nonbonding electron pairs on carbon atoms are very rare. Another useful rule of a similar nature is that if a species contains several atoms of one kind and one of another, the atoms of the same kind will assume equivalent positions in the species. In SO_4^{2-}, for example, the four O atoms are all equivalent, and are bonded to the S atom and not to one another.

Resonance structures are reasonably common. For resonance to occur, however, the atomic arrangement must remain fixed for two or more possible electronic structures. For CH_2O there are no resonance structures.

A. Using the procedure we have outlined, construct and report on models of the molecules and ions listed here and/or other species assigned by your instructor. Draw the complete Lewis structure for each molecule, showing nonbonding as well as bonding electrons. Given the structure, describe the geometry of the molecule or ion, and state whether the species is polar. Finally, draw the Lewis structures of any likely isomers or resonance forms.

CH_4	H_3O^+	N_2	C_2H_2	SCN^-
CH_2Cl_2	HF	P_4	SO_2	NO_3^-
CH_4O	NH_3	C_2H_4	SO_4^{2-}	HNO_3
H_2O	H_2O_2	$C_2H_2Br_2$	CO_2	$C_2H_4Cl_2$

B. Assuming that stability requires that each atom obey the octet rule, predict the stability of the following species:

$$PCl_3 \quad H_3O \quad CH_2 \quad CO$$

C. When you have completed parts A and B, see your laboratory instructor, who will check your results and assign you a set of unknown species. Working now by yourself, assemble models for each species as in the previous section, and report on the geometry and bonding in each of the unknown species on the basis of the model you construct. Also consider and report the polarity and the Lewis structures of any isomers and resonance forms for each species.

Experiment 13

Observations and Conclusions: The Geometrical Structure of Molecules

A. Species	Lewis structure	Molecular geometry	Polar?	Isomers or resonance structures
CH_4				
CH_2Cl_2				
CH_4O				
H_2O				
H_3O^+				
HF				
NH_3				
H_2O_2				
N_2				
P_4				
C_2H_4				

A. Species | **Lewis structure** | **Molecular geometry** | **Polar?** | **Isomers or resonance structures**

$C_2H_2Br_2$

C_2H_2

SO_2

SO_4^{2-}

CO_2

SCN^-

NO_3^-

HNO_3

$C_2H_4Cl_2$

B. Stability predicted for PCl_3 _____ H_3O _____ CH_2 _____ CO _____

C. Unknowns

_____

_____

_____

_____

Experiment 13

Advance Study Assignment: The Geometrical Structure of Molecules

You are asked by your instructor to construct a model of the CH_2Cl_2 molecule. Being of a conservative nature, you proceed as directed in the section on Experimental Procedure.

1. First you need to find the number of valence electrons in CH_2Cl_2. The number of valence electrons in an atom of an element is equal to the group number of that element in the Periodic Table.

    C is in Group _____ H is in Group _____ Cl is in Group _____

    In CH_2Cl_2 there is a total of _____ valence electrons.

2. The model consists of balls and sticks. What kind of ball should you select for the C atom? _____ the H atoms? _____ the Cl atom? _____ The electrons in the molecule are paired, and each stick represents an electron pair. How many sticks do you need? _____

3. Assemble a skeleton structure for the molecule, connecting the balls with sticks into one unit. Use the rule that C atoms form four bonds, whereas Cl atoms usually do not. Draw a sketch of the skeleton below:

4. How many sticks did you need to make the skeleton structure?_____ How many sticks are left over? _____ If your model is to obey the octet rule, each ball must have four sticks in it (except for hydrogen atom balls, which need only one). (Each atom in an octet rule species is surrounded by four pairs of electrons.) How many holes remain to be filled?_____ Fill them with the remaining sticks, which represent nonbonding electron pairs. Draw the complete Lewis structure for CH_2Cl_2 using lines for bonds and pairs of dots for nonbonding electrons.

5. Describe the geometry of the model, which is that of CH_2Cl_2. _____ Is the CH_2Cl_2 molecule polar? _____ Why?

    Would you expect CH_2Cl_2 to have any isomeric forms? _____ Explain your reasoning.

6. Would CH_2Cl_2 have any resonance structures? _____ If so, draw them below.

10

Heat Effects and Calorimetry

Heat is a form of energy, sometimes called thermal energy, that can pass spontaneously from an object at a high temperature to an object at a lower temperature. If the two objects are in contact, they will, given sufficient time, both reach the same temperature.

Heat flow is ordinarily measured in a device called a calorimeter. A calorimeter is simply a container with insulating walls, made so that essentially no heat is exchanged between the contents of the calorimeter and the surroundings. Within the calorimeter chemical reactions may occur or heat may pass from one part of the contents to another, but no heat flows into or out of the calorimeter from or to the surroundings.

A. Specific Heat

When heat flows into a substance, the temperature of that substance will increase. The quantity of heat q required to cause a temperature change Δt of any substance is proportional to the mass m of the substance and the temperature change, as shown in Equation 1. The proportionality constant is called the specific heat, $S.H.$, of that substance.

$$q = (\text{specific heat}) \times m \times \Delta t = S.H. \times m \times \Delta t \qquad (1)$$

The specific heat can be considered to be the amount of heat required to raise the temperature of one gram of the substance by 1°C (if you make m and Δt in Equation 1 both equal to 1, then q will equal $S.H.$). Amounts of heat are measured in either joules or calories. To raise the temperature of 1 g of water by 1°C, 4.18 joules of heat must be added to the water. The specific heat of water is therefore 4.18 joules/g°C. Since 4.18 joules equals 1 calorie, we can also say that the specific heat of water is 1 calorie/g°C. Ordinarily heat flow into or out of a substance is determined by the effect that that flow has on a known amount of water. Because water plays such an important role in these measurements, the calorie, which was the unit of heat most commonly used until recently, was actually defined to be equal to the specific heat of water.

The specific heat of a metal can readily be measured in a calorimeter. A weighed amount of metal is heated to some known temperature and is then quickly poured into a calorimeter that contains a measured amount of water at a known temperature. Heat flows from the metal to the water, and the two equilibrate at some temperature between the initial temperatures of the metal and the water.

Assuming that no heat is lost from the calorimeter to the surroundings, and that a negligible amount of heat is absorbed by the calorimeter walls, the amount of heat that flows from the metal as it cools is equal to the amount of heat absorbed by the water.

In thermodynamic terms, the heat flow for the metal is equal in magnitude but opposite in direction, and hence in sign, to that for the water. For the heat flow q,

$$q_{H_2O} = -q_{metal} \qquad (2)$$

If we now express heat flow in terms of Equation 1 for both the water and the metal M, we get

$$q_{H_2O} = S.H._{H_2O} m_{H_2O} \Delta t_{H_2O} = -S.H._M m_M \Delta t_M \qquad (3)$$

In this experiment we measure the masses of water and metal and their initial and final temperatures. (Note that $\Delta t_M < 0$ and $\Delta t_{H_2O} > 0$, since $\Delta t = t_{final} - t_{initial}$.) Given the specific heat of water, we can find the positive specific heat of the metal by Equation 3. We will use this procedure to obtain the specific heat of an unknown metal.

The specific heat of a metal is related in a simple way to its molar mass. Dulong and Petit discovered many years ago that about 25 joules were required to raise the temperature of one mole of many metals by 1°C. This relation, shown in Equation 4, is known as the Law of Dulong and Petit:

$$MM \cong \frac{25}{S.H.(J/g°C)} \qquad (4)$$

where MM is the molar mass of the metal. Once the specific heat of the metal is known, the approximate molar mass can be calculated by Equation 4. The Law of Dulong and Petit was one of the few rules available to early chemists in their studies of molar masses.

B. Heat of Reaction

When a chemical reaction occurs in water solution, the situation is similar to that which is present when a hot metal sample is put into water. With such a reaction there is an exchange of heat between the reaction mixture and the solvent, water. As in the specific heat experiment, the heat flow for the reaction mixture is equal in magnitude but opposite in sign to that for the water. The heat flow associated with the reaction mixture is also equal to the enthalpy change, ΔH, for the reaction, so we obtain the equation

$$q_{reaction} = \Delta H_{reaction} = -q_{H_2O} \qquad (5)$$

By measuring the mass of the water used as solvent, and by observing the temperature change that the water undergoes, we can find q_{H_2O} by Equation 1 and ΔH by Equation 5. If the temperature of the water goes up, heat has been *given off* by the reaction mixture, so the reaction is *exo*thermic; q_{H_2O} is *positive* and ΔH is *negative*. If the temperature of the water goes down, the reaction mixture has *absorbed heat from* the water and the reaction is *endo*thermic. In this case q_{H_2O} is *negative* and ΔH is *positive*. Both exothermic and endothermic reactions are observed.

One of the simplest reactions that can be studied in solution occurs when a solid is dissolved in water. As an example of such a reaction note the solution of NaOH in water:

$$NaOH(s) \rightarrow Na^+(aq) + OH^-(aq); \quad \Delta H = \Delta H_{solution} \qquad (6)$$

When this reaction occurs, the temperature of the solution becomes much higher than that of the NaOH and water that were used. If we dissolve a known amount of NaOH in a measured amount of water in a calorimeter, and measure the temperature change that occurs, we can use Equation 1 to find q_{H_2O} for the reaction and use Equation 5 to obtain ΔH. Noting that ΔH is directly proportional to the amount of NaOH used, we can easily calculate $\Delta H_{solution}$ for either a gram or a mole of NaOH. In the second part of this experiment you will measure $\Delta H_{solution}$ for an unknown ionic solid.

Chemical reactions often occur when solutions are mixed. A precipitate may form, in a reaction opposite in direction to that in Equation 6. A very common reaction is that of neutralization, which occurs when an acidic solution is mixed with one that is basic. In the last part of this experiment you will measure the heat effect when a solution of HCl, hydrochloric acid, is mixed with one containing NaOH, sodium hydroxide, which is basic. The heat effect is quite large, and is the result of the reaction between H^+ ions in the HCl solution with OH^- ions in the NaOH solution:

$$H^+(aq) + OH^-(aq) \rightarrow H_2O \quad \Delta H = \Delta H_{neutralization} \qquad (7)$$

Experimental Procedure

A. Specific Heat

From the stockroom obtain a calorimeter, a digital or sensitive mercury-in-glass thermometer, a sample of metal in a large stoppered test tube, and a sample of unknown solid. (The thermometer is expensive, so be careful when handling it.)

The calorimeter consists of two nested expanded polystyrene coffee cups fitted with a styrofoam cover. There are two holes in the cover for a thermometer and a glass stirring rod with a loop bend on one end. Assemble the experimental setup as shown in Figure 14.1.

Fill a 400-cm³ beaker two thirds full of water and begin heating it to boiling. While the water is heating, weigh your sample of unknown metal in the large stoppered test tube to the nearest 0.1 g on a top-loading or triple-beam balance. Pour the metal into a dry container and weigh the empty test tube and stopper. Replace the metal in the test tube and put the *loosely* stoppered tube into the hot water in the beaker. The water level in the beaker should be high enough so that the top of the metal is below the water surface. Continue heating the metal in the water for at least 10 minutes after the water begins to boil to ensure that the metal attains the temperature of the boiling water. Add water as necessary to maintain the water level.

While the water is boiling, weigh the calorimeter to 0.1 g. Place about 40 cm³ of water in the calorimeter and weigh again. Insert the stirrer and thermometer into the cover and put it on the calorimeter. The thermometer bulb should be completely under the water.

Measure the temperature of the water in the calorimeter to 0.1°C. Take the test tube out of the beaker of boiling water, remove the stopper, and pour the metal into the water in the calorimeter. Be careful that no water adhering to the outside of the test tube runs into the calorimeter when you are pouring the metal. Replace the calorimeter cover and agitate the water as best you can with the glass stirrer. Record to 0.1°C the maximum temperature reached by the water. Repeat the experiment, using about 50 cm³ of water in the calorimeter.

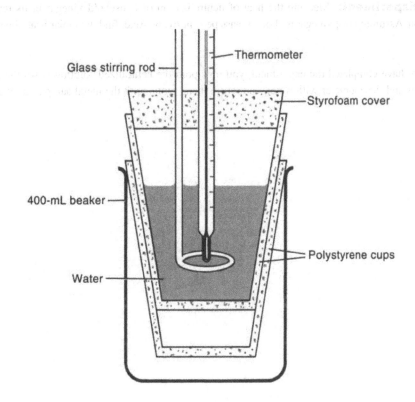

Be sure to dry your metal before reusing it; this can be done by heating the metal briefly in the test tube in boiling water and then pouring the metal onto a paper towel to drain. You can dry the hot test tube with a little compressed air.

B. Heat of Solution

Place about 50 cm³ of distilled water in the calorimeter and weigh as in the previous procedure. Measure the temperature of the water to 0.1°C. The temperature should be within a degree or two of room temperature. In a small beaker weigh out about 5 g of the solid compound assigned to you. Make the weighing of the beaker and of the beaker plus solid to 0.1 g. Add the compound to the calorimeter. Stirring continuously and occasionally swirling the calorimeter, determine to 0.1°C the maximum or minimum temperature reached as the solid dissolves. Check to make sure that *all* the solid dissolved. A temperature change of at least 5 degrees should be obtained in this experiment. If necessary, repeat the experiment, increasing the amount of solid used.

   DISPOSAL OF REACTION PRODUCTS.   Dispose of the solution in Part B as directed by your instructor.

C. Heat of Neutralization

Rinse out your calorimeter with distilled water, pouring the rinse into the sink. In a graduated cylinder, measure out 25 cm³ of 1.00 M HCl; pour that solution into the calorimeter. Rinse out the cylinder with distilled water, and measure out 25 cm³ of 1.00 M NaOH; pour that solution into a dry 50-mL beaker. Measure the temperature of the acid and of the base to ±0.1°C, making sure to rinse and dry your thermometer before immersing it in the solutions. Put the thermometer back in the calorimeter cover. Pour the NaOH solution into the HCl solution and put on the cover of the calorimeter. Stir the reaction mixture, and record the maximum temperature that is reached by the neutralized solution.

Optional Experiment: Measure the heat of neutralization of household vinegar in its reaction with NaOH solution. Assuming that vinegar is about 5 mass percent Acetic Acid, find the molar heat of neutralization of that acid.

   When you have completed the experiment, you may pour the neutralized solution down the sink. Rinse the calorimeter and thermometer with water, and return them, along with the metal sample to the stockroom.

Experiment 14

Data and Calculations: Heat Effects and Calorimetry

A. Specific Heat

	Trial 1	Trial 2
Mass of stoppered test tube plus metal	_____ g →	_____ g
Mass of test tube and stopper	_____ g →	_____ g
Mass of calorimeter	_____ g →	_____ g
Mass of calorimeter and water	_____ g	_____ g
Mass of water	_____ g	_____ g
Mass of metal	_____ g →	_____ g
Initial temperature of water in calorimeter	_____ °C	_____ °C
Initial temperature of metal (assume 100°C unless directed to do otherwise)	_____ °C →	_____ °C
Equilibrium temperature of metal and water in calorimeter	_____ °C	_____ °C
Δt_{water} $(t_{final} - t_{initial})$	_____ °C	_____ °C
Δt_{metal}	_____ °C	_____ °C
q_{H_2O}	_____ J	_____ J
Specific heat of the metal (Eq. 3)	_____ J/g°C	_____ J/g°C
Approximate molar mass of metal	_____	_____
Unknown no.		_____

B. Heat of Solution

Mass of calorimeter plus water	_____ g
Mass of beaker	_____ g

Mass of beaker plus solid   _____ g

Mass of water, m_{H_2O} _____ g

Mass of solid, m_s _____ g

Original temperature   _____ °C

Final temperature   _____ °C

q_{H_2O} for the reaction (Eq. 1) (S.H. = 4.18 J/g°C)   _____ joules

ΔH for the reaction (Eq. 5)   _____ joules

The quantity you have just calculated is approximately* equal to the heat of solution of your sample. Calculate the heat of solution per gram of solid sample.

$$\Delta H_{solution} = \underline{\hspace{2cm}} \text{ joules/g}$$

The solution reaction is endothermic exothermic. (Underline correct answer.) Give your reasoning.

Solid unknown no. _____

Optional Formula of compound used (if furnished) _____ Molar mass _____ g

Heat of solution per mole of compound   _____ kJ

C. Heat of Neutralization

Original temperature of HCl solution   _____ °C

Original temperature of NaOH solution   _____ °C

Final temperature of neutralized mixture   _____ °C

Change in temperature. Δt (take average of the original temperatures of HCl and NaOH)   _____ °C

q_{H_2O} (assume 50 g H_2O are present)   _____ J

ΔH for the neutralization reaction   _____ J

ΔH per mole of H^+ and OH^- ions reacting   _____ kJ

Experiment 14

Advance Study Assignment: Heat Effects and Calorimetry

1. A metal sample weighing 71.9 g and at a temperature of 100.0°C was placed in 41.0 g of water in a calorimeter at 24.5°C. At equilibrium the temperature of the water and metal was 35.0°C.

   a. What was Δt for the water? ($\Delta t = t_{final} - t_{initial}$)

   _____ °C

   b. What was Δt for the metal?

   _____ °C

   c. How much heat flowed into the water?

   _____ joules

   d. Taking the specific heat of water to be 4.18 J/g°C, calculate the specific heat of the metal, using Equation 3.

   _____ joules/g°C

   e. What is the approximate molar mass of the metal? (Use Eq. 4.)

   _____ g

2. When 2.0 g of NaOH were dissolved in 49.0 g water in a calorimeter at 24.0°C, the temperature of the solution went up to 34.5°C.

   a. Is this solution reaction exothermic? _____ Why?

   b. Calculate q_{H_2O}, using Equation 1.

   _____ joules

   c. Find ΔH for the reaction as it occurred in the calorimeter (Eq. 5).

   $\Delta H =$ _____ joules

d. Find ΔH for the solution of 1.00 g NaOH in water.

$\Delta H = $ _____ joules/g

e. Find ΔH for the solution of 1 mole NaOH in water.

$\Delta H = $ _____ joules/mole

f. Given that NaOH exists as Na^+ and OH^- ions in solution, write the equation for the reaction that occurs when NaOH is dissolved in water.

g. Given the following heats of formation, ΔH_f, in kJ per mole, as obtained from a table of ΔH_f data, calculate ΔH for the reaction in Part f. Compare your answer with the result you obtained in Part e.
    NaOH(s), -425.6; Na^+(aq), -240.1; OH^-(aq), -230.0

$\Delta H = $ _____ kJ

11

The Standardization of a Basic Solution and the Determination of the Molar Mass of an Acid

When a solution of a strong acid is mixed with a solution of a strong base, a chemical reaction occurs that can be represented by the following net ionic equation:

$$H^+(aq) + OH^-(aq) \rightarrow H_2O$$

This is called a neutralization reaction, and chemists use it extensively to change the acidic or basic properties of solutions. The equilibrium constant for the reaction is about 10^{14} at room temperature, so that the reaction can be considered to proceed completely to the right, using up whichever of the ions is present in the lesser amount and leaving the solution either acidic or basic, depending on whether H^+ or OH^- ion was in excess.

Since the reaction is essentially quantitative, it can be used to determine the concentrations of acidic or basic solutions. A frequently used procedure involves the titration of an acid with a base. In the titration, a basic solution is added from a buret to a measured volume of acid solution until the number of moles of OH^- ion added is just equal to the number of moles of H^+ ion present in the acid. At that point the volume of basic solution that has been added is measured.

Recalling the definition of the molarity, M_A, of species A, we have

$$M_A = \frac{\text{no. moles of A}}{\text{no. liters of solution, } V} \quad \text{or} \quad \text{no. moles of A} = M_A \times V \tag{1}$$

At the end point of the titration,

$$\text{no. moles } H^+ \text{ originally present} = \text{no. moles } OH^- \text{ added} \tag{2}$$

So, by Equation 1,

$$M_{H^+} \times V_{acid} = M_{OH^-} \times V_{base} \tag{3}$$

Therefore, if the molarity of either the H^+ or the OH^- ion in its solution is known, the molarity of the other ion can be found from the titration.

The equivalence point or end point in the titration is determined by using a chemical, called an indicator, that changes color at the proper point. The indicators used in acid-base titrations are weak organic acids or bases that change color when they are neutralized. One of the most common indicators is phenolphthalein, which is colorless in acid solutions but becomes red when the pH of the solution becomes 9 or higher.

When a solution of a strong acid is titrated with a solution of a strong base, the pH at the end point will be about 7. At the end point a drop of acid or base added to the solution will change its pH by several pH units, so that phenolphthalein can be used as an indicator in such titrations. If a weak acid is titrated with a strong base, the pH at the equivalence point is somewhat higher than 7, perhaps 8 or 9, and phenolphthalein is still a very satisfactory indicator. If, however, a solution of a weak base such as ammonia is titrated with a strong acid, the pH will be a unit or two less than 7 at the end point, and phenolphthalein will not be as good an indicator for that titration as, for example, methyl red, whose color changes from red to yellow as the pH changes from about 4 to 6. Ordinarily, indicators will be chosen so that their color change occurs at about the pH at the equivalence point of a given acid-base titration.

In this experiment you will determine the molarity of OH⁻ ion in an NaOH solution by titrating that solution against a standardized solution of HCl. Since in these solutions one mole of acid in solution furnishes one mole of H⁺ ion and one mole of base produces one mole of OH⁻ ion, $M_{HCl} = M_{H^+}$ in the acid solution, and $M_{NaOH} = M_{OH^-}$ in the basic solution. Therefore, the titration will allow you to find M_{NaOH} as well as M_{OH^-}.

In the second part of this experiment you will use your standardized NaOH solution to titrate a sample of a pure solid organic or inorganic acid. By titrating a weighed sample of the unknown acid with your standardized NaOH solution you can, by Equation 2, find the number of moles H⁺ ion that your sample can furnish.

If your acid has one acidic hydrogen atom in the molecule, with formula HB, then the number of moles of acid will equal the number of moles of H⁺ that react during the titration. The molar mass of the acid, MM, will equal the number of grams of acid that contain one mole of H⁺ ion.

$$MM = \frac{\text{no. grams of acid}}{\text{no. moles of H}^+ \text{ ion furnished}} \qquad (4)$$

Many acids release one mole of H⁺ ion per mole of acid on titration with NaOH solution. Such acids are called monoprotic. Acetic acid, $HC_2H_3O_2$, is a classic example of a monoprotic acid (only the first H atom in the formula is acidic). Like all organic acids, acetic acid is weak, in that it only ionizes to a small extent in water solution.

Some acids contain more than one acidic hydrogen atom in the molecule, and would have the general formula H_2B or H_3B. Sulfurous acid, H_2SO_3, is an example of an inorganic diprotic acid. Maleic acid, $H_2C_4H_2O_4$, is a diprotic organic acid. If we should titrate samples of these acids with a solution of a strong base like NaOH, it would take two moles of OH⁻ ion, or two moles of NaOH, to neutralize one mole of acid, since each mole of acid would release two moles of H⁺ ion. If you don't know the formula of an acid, you can't be sure it is monoprotic, so you can only calculate the mass of acid that will react with one mole of OH⁻ ion. That mass will contain one mole of H⁺ ion and is called the *equivalent* mass of the acid. The molar mass and the equivalent mass are related by simple equations. Since sulfurous acid gives up two moles of H⁺ ion in titration with NaOH, the molar mass must equal *twice* the equivalent mass, which gives up one mole of H⁺ ion.

To simplify matters, in this experiment we will only use monoprotic acids, with formula HB, so one mole of acid will react with one mole of NaOH, and you can find the molar mass of your acid by Equation 4.

Experimental Procedure

WEAR YOUR SAFETY GLASSES WHILE
PERFORMING THIS EXPERIMENT

Note: This experiment is relatively long unless you know precisely what to do. Study the experiment carefully before coming to class, so that you don't have to spend a lot of time finding out what the experiment is all about.

Obtain two burets and a sample of solid unknown acid from the stockroom.

A. Standardization of NaOH Solution

Into a small graduated cylinder draw about 7 mL of the stock 6 M NaOH solution provided in the laboratory and dilute to about 400 mL with distilled water in a 500-mL Florence flask. Stopper the flask tightly and mix the solution thoroughly at intervals over a period of at least 15 minutes before using the solution.

Draw into a clean, *dry* 125-mL Erlenmeyer flask about 75 mL of standardized HCl solution (about 0.1 M) from the stock solution on the reagent shelf. This amount should provide all the standard acid you will need; do not waste it. Record the molarity of the HCl.

Prepare for the titration by using the procedure described in Experiment 7 and Appendix IV. The purpose of this procedure is to make sure that the solution in each buret has the same molarity as it has in the container from which it was poured. Clean the two burets and rinse them with distilled water. Then rinse one buret three times with a few milliliters of the HCl solution, in each case thoroughly wetting the walls of the buret with the solution and then letting it out through the stopcock. Fill the buret with HCl; open the stopcock momentarily to fill the tip. Proceed to clean and fill the other buret with your NaOH solution in a similar manner. Put the acid buret, A, on the left side of your buret clamp, and the base buret, B, on the right side. Check to see that your burets do not leak and that there are no air bubbles in either buret tip. Read and record the levels in the two burets to 0.02 mL.

Draw about 25 mL of the HCl solution from the buret into a clean 250-mL Erlenmeyer flask; add to the flask about 25 mL of distilled H_2O and two or three drops of phenolphthalein indicator solution. Place a white sheet of paper under the flask to aid in the detection of any color change. Add the NaOH solution intermittently from its buret to the solution in the flask, noting the pink phenolphthalein color that appears and disappears as the drops hit the liquid and are mixed with it. Swirl the liquid in the flask gently and continuously as you add the NaOH solution. When the pink color begins to persist, slow down the rate of addition of NaOH. In the final stages of the titration add the NaOH drop by drop until the entire solution just turns a pale pink color that will persist for about 30 seconds. If you go past the end point and obtain a red solution, add a few drops of the HCl solution to remove the color, and then add NaOH a drop at a time until the pink color persists. Carefully record the *final* readings on the HCl and NaOH burets.

To the 250-mL Erlenmeyer flask containing the titrated solution, add about 10 mL more of the standard HCl solution. Titrate this as before with the NaOH to an end point, and carefully record both buret readings once again. To this solution add about 10 mL more HCl and titrate a third time with NaOH.

You have now completed three titrations, with *total* HCl volumes of about 25, 35, and 45 mL. Using Equation 3, calculate the molarity of your base, M_{OH^-} for each of the three titrations. In each case, use the *total volumes* of acid and base that were added up to that point. At least two of these molarities should agree to within 1%. If they do, proceed to the next part of the experiment. If they do not, repeat these titrations until two calculated molarities do agree.

B. Determination of the Molar Mass of an Acid

Weigh the vial containing your solid acid on the analytical balance to ±0.0001 g. Carefully pour out about half the sample into a clean but not necessarily dry 250-mL Erlenmeyer flask. Weigh the vial accurately. Add about 50 mL of distilled water and two or three drops of phenolphthalein to the flask. The acid may be relatively insoluble, so don't worry if it doesn't all dissolve.

Fill your NaOH buret with the (now standardized) NaOH solution. Add the standard HCl to your HCl buret until it is about half full. Read both levels carefully and record them.

Titrate the solution of the solid acid with NaOH. As the acid is neutralized it will tend to dissolve in the solution. If the acid appears to be relatively insoluble, add NaOH until the pink color persists, and then swirl to dissolve the solid. If the solid still will not dissolve, and the solution remains pink, add 25 mL of ethanol to increase the solubility. If you go past the end point, add HCl as necessary. The final pink end point should appear on addition of one drop of NaOH. Record the final levels in the NaOH and HCl burets.

Pour the rest of your acid sample into a clean 250-mL Erlenmeyer flask, and weigh the vial accurately. Titrate this sample of acid as before with the NaOH and HCl solutions.

If you use HCl in these titrations, and you probably will, the calculations needed are a bit more complicated than in the standardization of the NaOH solution. To find the number of moles of H^+ ion in the solid acid, you must subtract the number of moles of HCl used from the number of moles of NaOH. For a back-titration, which is what we have in this case:

$$\text{no. moles } H^+ \text{ in solid acid} = \text{no. moles } OH^- \text{ in NaOH soln.} - \text{no. moles } H^+ \text{ in HCl soln.} \qquad (5)$$

For volumes in milliliters, this equation takes the form:

$$\text{no. moles } H^+ \text{ in solid acid} = \frac{M_{NaOH} \times V_{NaOH}}{1000} - \frac{M_{HCl} \times V_{HCl}}{1000} \qquad (6)$$

C. Determination of K_a for the Unknown Acid   Optional

Your instructor will tell you in advance if you are to do this part of the experiment. If you do this part, you will need to **save** one of the titrated solutions from Part B. Given that titrated solution, it is easy to prepare one in which [H⁺] ion is equal to K_a for your acid.

A weak acid, HB, dissociates in water solution according to the equation:

$$HB(aq) \rightleftharpoons H^+(aq) + B^-(aq) \tag{7}$$

The equilibrium constant for this reaction is called the acid dissociation constant for HB and is given the symbol K_a. A solution of HB will obey the equilibrium condition given by the equation:

$$K_a = \frac{[H^+][B^-]}{[HB]} \tag{8}$$

Your acid is a weak acid, so will follow both of the above equations.

In the titration in Part B you converted a solution of HB into one containing NaB by adding NaOH to the end point. If, to your titrated solution, you now add some HCl, it will convert some of the B⁻ ions in the NaB solution back to HB. If the number of moles of HCl you add equals *one-half* the number of moles of H⁺ in your original sample, then *half* of the B⁻ ions will be converted to HB. In the resulting solution, [HB] will equal [B⁻], and so, by Equation 8,

$$K_a = [H^+] \quad \text{in the half-neutralized solution} \tag{9}$$

Using the approach we have outlined, use your titrated solution to prepare one in which [HB] equals [B⁻]. Measure the pH of that solution with a pH meter, using the procedure described in Appendix IV. From that value, calculate K_a for your unknown acid.

Optional Experiment: Using the procedure in this experiment, find the mass percent acetic acid in household vinegar. What is the molarity of the acid?

   DISPOSAL OF REACTION PRODUCTS.   The reaction products in this experiment may be diluted and poured down the drain.

Experiment 24

Data and Calculations: The Standardization of a Basic Solution and the Determination of the Molar Mass of an Acid

A. Standardization of NaOH Solution

	Trial 1	**Trial 2**	**Trial 3**
Initial reading, HCl buret	_____ mL		
Final reading, HCl buret	_____ mL	_____ mL	_____ mL
Initial reading, NaOH buret	_____ mL		
Final reading, NaOH buret	_____ mL	_____ mL	_____ mL

B. Determination of the Molar Mass of an Unknown Acid

Mass of vial plus contents	_____ g
Mass of vial plus contents less Sample 1	_____ g
Mass less Sample 2	_____ g

	Trial 1	**Trial 2**
Initial reading, NaOH buret	_____ mL	_____ mL
Final reading, NaOH buret	_____ mL	_____ mL
Initial reading, HCl buret	_____ mL	_____ mL
Final reading, HCl buret	_____ mL	_____ mL

Processing the Data

A. Standardization of NaOH Solution

	Trial 1	**Trial 2**	**Trial 3**
Total volume HCl	_____ mL	_____ mL	_____ mL
Total volume NaOH	_____ mL	_____ mL	_____ mL

Molarity, M_{HCl}, of standardized HCl   _____ M

Molarity, M_{H^+}, in standardized HCl   _____ M

By Equation 3,

$$M_{H^+} \times V_{acid} = M_{OH^-} \times V_{base} \quad \text{or} \quad M_{OH^-} = M_{H^+} \times \frac{V_{HCl}}{V_{NaOH}} \qquad \textbf{(3)}$$

Use Equation 3 to find the molarity, M_{OH^-}, of the NaOH solution. Note that the volumes do not need to be converted to liters, since we use the volume ratio.

	Trial 1	**Trial 2**	**Trial 3**

M_{OH^-} _____ M   _____ M   _____ M   (should agree within 1%)

The molarity of the NaOH will equal M_{OH^-}, since one mole NaOH $\rightarrow$ one mole OH^-.

Average molarity of NaOH solution, M_{NaOH} _____ M

B. Determination of the Molar Mass of the Unknown Acid

	Trial 1	**Trial 2**
Mass of sample	_____ g	_____ g
Volume NaOH used	_____ mL	_____ mL
No. moles NaOH $= \dfrac{V_{NaOH} \times M_{NaOH}}{1000}$	_____	_____
Volume HCl used	_____ mL	_____ mL
No. moles HCl $= \dfrac{V_{HCl} \times M_{HCl}}{1000}$	_____	_____
No. moles H^+ in sample (use Eq. 5)	_____	_____
$MM = \dfrac{\text{no. grams acid}}{\text{no. moles } H^+}$	_____ g	_____ g

Unknown no.   _____

C. Determination of K_a for the Unknown Acid Optional

No. moles H^+ in sample (from Part B)   _____

No. moles HCl to be added _____   Volume of HCl added _____ mL

pH of half-neutralized solution _____ K_a of acid _____

Experiment 24

Advance Study Assignment: Molar Mass of an Acid

1. 7.0 mL of 6.0 M NaOH are diluted with water to a volume of 400 mL. You are asked to find the molarity of the resulting solution.

   a. First find out how many moles of NaOH there are in 7.0 mL of 6.0 M NaOH. Use Equation 1. Note that the volume must be in liters.

   _____ moles

   b. Since the total number of moles of NaOH is not changed on dilution, the molarity after dilution can also be found by Equation 1, using the final volume of the solution. Calculate that molarity.

   _____ M

2. In an acid-base titration, 22.13 mL of an NaOH solution are needed to neutralize 24.65 mL of a 0.1094 M HCl solution. To find the molarity of the NaOH solution, we can use the following procedure:

   a. First note the value of M_{H^+} in the HCl solution.

   _____ M

   b. Find M_{OH^-} in the NaOH solution. (Use Eq. 3.)

   _____ M

   c. Obtain M_{NaOH} from M_{OH^-}.

   _____ M

3. A 0.2678 g sample of an unknown acid requires 27.21 mL of 0.1164 M NaOH for neutralization to a phenolphthalein end point. There are 0.35 mL of 0.1012 M HCl used for back-titration.

   a. How many moles of OH^- are used? How many moles of H^+ from HCl?

   _____ moles OH^- _____ moles H^+

   b. How many moles of H^+ are there in the solid acid? (Use Eq. 5.)

   _____ moles H^+ in solid

   c. What is the molar mass of the unknown acid? (Use Eq. 4.)

   _____ g

Experiment 24

Advance Study Assignment: Molar Mass of an Acid

Appendix I

Vapor Pressure of Water

Temperature °C	Pressure mm Hg	Temperature °C	Pressure mm Hg
0	4.6	26	25.2
1	4.9	27	26.7
2	5.3	28	28.3
3	5.7	29	30.0
4	6.1	30	31.8
5	6.5	31	33.7
6	7.0	32	35.7
7	7.5	33	37.7
8	8.0	34	39.9
9	8.6	35	42.2
10	9.2	40	55.3
11	9.8	45	71.9
12	10.5	50	92.5
13	11.2	55	118.0
14	12.0	60	149.4
15	12.8	65	187.5
16	13.6	70	233.7
17	14.5	75	289.1
18	15.5	80	355.1
19	16.5	85	433.6
20	17.5	90	525.8
21	18.7	95	633.9
22	19.8	97	682.1
23	21.1	99	733.2
24	22.4	100	760.0
25	23.8	101	787.6

To convert mm Hg to kPa, multiply the entry in the table by 0.1333.

$$1 \text{ mm Hg} = 0.1333 \text{ kPa}$$
$$= 13.57 \text{ mm } H_2O$$

Vapor Pressure of Water

Appendix II

Summary of Solubility Properties of Ions and Solids

	Cl⁻, Br⁻ I⁻, SCN⁻	SO_4^{2-}	CrO_4^{2-}	PO_4^{3-}	$C_2O_4^{2-*}$	CO_3^{2-}
Na^+, K^+, NH_4^+	S	S	S	S	S	S
Ba^{2+}	S	I	A	A⁻	A	A⁻
Ca^{2+}	S	S⁻	S	A⁻	A	A⁻
Mg^{2+}	S	S	S	A⁻	A⁻	A⁻
Fe^{3+} (yellow)	S*	S	A⁻	A	S	D, A⁻
Cr^{3+} (blue-gray)	S	S	A⁻	A	S	A⁻
Al^{3+}	S	S	A⁻, C	A, C	A⁻, C	D, A⁻, C
Ni^{2+} (green)	S	S	S	A⁻, C	A, C	A⁻, C
Co^{2+} (pink)	S	S	A⁻	A⁻	A⁻	A⁻
Zn^{2+}	S	S	A⁻, C	A⁻, C	A⁻, C	A⁻, C
Mn^{2+} (pale pink)	S	S	S	A⁻	A⁻	A⁻
Cu^{2+} (blue)	S*	S	A⁻, C	A⁻, C	A, C	A⁻, C
Cd^{2+}	S	S	A⁻, C	A⁻, C	A, C	A⁻, C
Bi^{3+}	A	A⁻	A	A	A	A⁻
Hg^{2+}	S*	S	A	A⁻	A	A⁻
Sn^{2+}, Sn^{4+}	A, C	A, C	A, C	A, C	A, C	A, C
Sb^{3+}	A, C	A, C	A, C	A, C	A⁻, C	A, C
Ag^+	C*	S⁻	A, C	A, C	A, C	A⁻, C
Pb^{2+}	C, HW	C	C	A, C	A, C	A⁻, C
Hg_2^{2+}	O⁺	A	A	A	O	A

	SO_3^{2-}	S^{2-}	O^{2-}, * OH^-	NO_3^-, ClO_3^- $C_2H_3O_2^-$, NO_2^-	Complexes
Na^+, K^+, N_4^+	S	S	S	S	—
Ba^{2+}	A	S	S^-	S	—
Ca^{2+}	A^-	D, A^-	S^-	S	—
Mg^{2+}	S	D, A^-	A^-	S	—
Fe^{3+} (yellow)	D, S	D, A^-	A^-	S	—
Cr^{3+} (blue-gray)	S	D, A^-	A^-	S	(OH^-)
Al^{3+}	A^-, C	D, A^-, C	A^-, C	S	OH^-
Ni^{2+} (green)	A^-	O	A^-, C	S	NH_3
Co^{2+} (pink)	A^-	O	A^-	S	*
Zn^{2+}	S	A^-, C	A^-, C	S	OH^-, NH_3
Mn^{2+} (pale pink)	S	A^-	A^-	S	—
Cu^{2+} (blue)	A^-, C	O	A^-, C	S	NH_3
Cd^{2+}	A^-, C	A	A^-, C	S	NH_3
Bi^{3+}	A	A^+, O	A^-	A^-	Cl^-
Hg^{2+}	D, O	O^+	A^-	S	—
Sn^{2+}, Sn^{4+}	A, C	A, C	A, C	A, C	OH^-, Cl^-
Sb^{3+}	A, C	A, C	A, C	A, C	OH^-, Cl^-
Ag^+	A, C	O	A^-, C	S	NH_3
Pb^{2+}	A, C	O	A^-, C	S	OH^-
Hg_2^{2+}	D, O	D, O^+	D, O	S	—

Key: S, soluble in water; no precipitate on mixing cation, 0.1 M, with anion, 1 M

    S^-, slightly soluble; tends to precipitate on mixing cation, 0.1 M, with anion, 1 M

    HW, soluble in hot water

    A^-, soluble in 1 M $HC_2H_3O_2$

    A, soluble in acid (6 M HCl or other nonprecipitating, nonoxidizing acid)

    A^+, soluble in 12 M HCl

    O, soluble in hot 6 M HNO_3

    O^+, soluble in aqua regia

    C, soluble in solution containing a good complexing ligand

    D, unstable, decomposes to a product with solubility as indicated

    I, insoluble in any common solvent

*Oxalates form many complex ions; oxides behave like hydroxides, but may be slow to dissolve; FeI_3 is unstable, decomposes to FeI_2 and I_2, CuI_2 is unstable, decomposes to CuI and I_2; AgBr and AgI do not dissolve in 6 M NH_3; HgI_2 is insoluble, but dissolves in excess I^-; Cr^{3+} and Co^{2+} can, under some conditions, form complexes with OH^- and NH_3 respectively.

Appendix IIA

Some Properties of the Cations in Groups I, II, and III

In the experiments on qualitative analysis we used the chemical and physical properties of the cations in Groups I, II, and III to separate the ions from each other. Here we have summarized most of the properties that were employed, and have included some others that help distinguish the cations we studied.

Lead ion, Pb^{2+}

Most lead compounds are insoluble in water; the only exceptions are $Pb(NO_3)_2$ and $Pb(C_2H_3O_2)_2$. Lead chloride is slightly soluble in water, much more soluble in hot water and in solutions where Cl^- concentration is high. Lead ion forms stable complexes with OH^-, $C_2H_3C_2^-$, and, to a lesser degree, with Cl^-. $PbSO_4$, an acid-insoluble sulfate, is white, and will dissolve in 6 M NaOH or in a concentrated solution of acetate ions by formation of the $Pb(OH)_3^-$ complex ion or $Pb(C_2H_3O_2)_2$, a species which does not appreciably dissociate in solution. $PbCrO_4$ is a bright yellow insoluble solid, often used as a confirmatory test for lead. It will dissolve in 6 M NaOH.

Silver ion, Ag^+

Most silver compounds are insoluble, the nitrate being just about the only exception. Silver acetate is slightly soluble. Silver chloride is white, insoluble in water but soluble in 6 M NH_3, due to formation of $Ag(NH_3)_2^+$ complex ion. These properties of AgCl are usually used in the confirmatory test for silver ion. They may also be used to test for the presence of Cl^- ion. In NaOH solution, Ag^+ precipitates as brown Ag_2O; it is soluble in nitric acid and in strong ammonia solutions.

Mercurous ion, Hg_2^{2+}

No salts of Hg(I) ion are soluble in water. Mercury(I) nitrate solutions always contain fairly high concentrations of HNO_3. Calomel, Hg_2Cl_2, precipitates on addition of HCl to any solution of Hg(I) ion. It is white, and essentially insoluble in all common solvents. If treated with 6 M NH_3, it turns black, due to a reaction to form Hg metal (black) and white, insoluble, $HgNH_2Cl$. This is the definitive test for Hg(I) ion.

Bismuth ion, Bi^{3+}

There are no water-soluble bismuth salts. The usual solution is highly acidic and contains $Bi(NO_3)_3$ or $BiCl_3$. Bismuth ion forms complexes with Cl^-, but not with OH^- or NH_3. $Bi(OH)_3$ is white and insoluble in water but soluble in strong acids. If you add a few drops of acidic $BiCl_3$ solution to water, a cloudy white precipitate of BiOCl forms. This is the usual test for Bi^{3+} ion. Another confirmatory test is to add 0.1 M $SnCl_2$ solution to $Bi(OH)_3$ in a strongly basic solution. Bi(III) will be reduced to black Bi metal.

Tin ion, Sn^{2+} or Sn^{4+}

No common tin compounds are water soluble. The usual 0.1 M tin solution contains $SnCl_2$ in 0.1 M HCl. Tin exists in either the +2, stannous, or +4, stannic, state. The Sn(II) species, particularly in basic solution, are good reducing agents, and can reduce Bi(III) and Hg(II) to the metals. Tin in either oxidation state forms many complex ions, and in solution is usually present as a complex ion. In basic solution Sn(II) exists as $Sn(OH)_4^{2-}$. In HCl the complex ion has the formula $SnCl_4^{2-}$. Tin compounds are white for the most part. SnS is brown, and its formation at pH 0.5 as in Experiment 37 is good evidence for the presence of tin. The usual confirmatory test is to add $SnCl_2$ in HCl to a solution of $HgCl_2$. The tin ion reduces the mercury(II) to form a precipitate of white Hg_2Cl_2, which may darken to gray as the reduction proceeds all the way to metallic mercury.

Antimony ion, Sb^{3+}

There are no common antimony salts that are soluble in water. Antimony salts are not easily dissolved. The usual solution of Sb(III) contains $SbCl_3$ in 3 M HCl, in which the antimony exists as $SbCl_6^{3-}$ complex ion. Antimony also forms a stable complex ion with hydroxide ion, $Sb(OH)_6^{3-}$. If a few drops of $SbCl_3$ in acidic solution are added to water, a white precipitate of SbOCl forms, very similar to that observed with $BiCl_3$ in solution. The confirmatory test for antimony is the formation of the characteristic red-orange sulfide on precipitation of Sb_2S_3 by addition of thioacetamide to antimony solution at pH 0.5. If the cation is Bi(III), the precipitate will be black Bi_2S_3.

Copper ion, Cu^{2+}

Copper(II) is colored in most of its compounds and solutions. The hydrated ion, $Cu(H_2O)_4^{2+}$ is blue, but in the presence of other complexing agents the Cu(II) ion may be green or dark blue. The sulfate, chloride, nitrate, and acetate are water soluble. If 6 M NH_3 is added to a solution of Cu(II) ion, light blue $Cu(OH)_2$ initially precipitates, but in excess reagent the hydroxide readily dissolves because of formation of the very characteristic dark blue copper(II) ammonia complex ion, $Cu(NH_3)_4^{2+}$. This ion serves as an excellent confirmatory test for the presence of copper. Copper ion is readily reduced to the metal by the more active metals, such as zinc.

Nickel ion, Ni^{2+}

Nickel is another of the colored cations. The chloride, nitrate, sulfate, and acetate are water soluble, and form green solutions containing the $Ni(H_2O)_6^{2+}$ complex ion. Nickel forms other complex ions, but the one most commonly observed is the blue $Ni(NH_3)_6^{2+}$ ion, formed on addition of 6 M NH_3 in excess to Ni(II) solutions. The color is much less intense than that of the copper ammonia complex ion. The usual test for the presence of nickel is the formation of a rose-red precipitate of nickel dimethylglyoxime on addition of that reagent to a slightly basic solution containing Ni^{2+}.

Iron(III) ion, Fe^{3+}

Iron has two common oxidation states, +2 and +3. Iron(II), called ferrous ion, is less commonly observed, since it is readily oxidized in air to the +3 state. We will limit our work to iron(III), or ferric, salts. Iron(III) salts are often colored, usually yellow in solution. The color is due to hydrolysis, since the $Fe(H_2O)_6^{3+}$ is itself colorless. The nitrate, chloride, and sulfate are water soluble, but tend to hydrolyze to form basic salts that may require a slightly acidic solution if they are to dissolve completely. Iron(III) hydroxide is very insoluble, and is the color of rust. Iron forms several complex ions, but the one most encountered in qualitative analysis is the $FeSCN^{2+}$ ion, which is dark red and often used in the confirmatory test for Fe(III) ion. Iron does not form complex ions with NH_3 or OH^-.

Chromium ion, Cr^{3+}

Chromium(III), as its name implies, is colored; it may be red-violet, green, or purple in solution. The chromium +3 ion usually exists as a complex which, unlike many complex ions, is sometimes slow to undergo ligand exchange. Chromium exists in two common oxidation states, +3 and +6. In the latter state it is found as the yellow chromate, CrO_4^{2-}, anion, or, if the solution is acidic, as the orange $Cr_2O_7^{2-}$ anion. The chloride, nitrate, sulfate, and acetate of Cr^{3+} are water soluble, but the solution process may be more rapid in acidic solution.

If excess 6 M NaOH is added to a solution of Cr(III), the initial hydroxide precipitate may dissolve and one may obtain a dark green solution due to formation of the $Cr(OH)_4^-$ complex ion. With 6 M NH_3 an insoluble grey hydroxide is initially formed, which may slowly dissolve to yield a pink or violet complex ion. Boiling a solution of either complex causes insoluble $Cr(OH)_3$ to re-form. In most qual schemes chromium is oxidized to the +6 state, where in acidic solution it forms a characteristic, but fleeting, deep blue solution on addition of H_2O_2. Where chromium(III) is the only cation in a sample, the formation of the green hydroxide complex is definitive. If this test is inconclusive, oxidize Cr(III) as in Experiment 38 and use the alternate confirmatory test.

Aluminum ion, Al^{3+}

Aluminum salts are typically white. The nitrate, chloride, sulfate, and acetate are water soluble, but do show a tendency to form basic salts if no acid is added. Aluminum forms a very stable hydroxide complex ion, $Al(OH)_4^-$, so it is difficult to precipitate the otherwise insoluble $Al(OH)_3$ by addition of NaOH to Al(III) solutions. That hydroxide will come down on addition of 6 M NH_3 to a solution containing Al^{3+} buffered with NH_4^+ ion. The precipitate is characteristically light, translucent, and gelatinous. A good confirmatory test for aluminum is to precipitate $Al(OH)_3$ from a solution of the sample. Dissolve the solid in very dilute acetic acid. Add a few drops of catechol violet reagent. If aluminum is present, a blue solution will form.

Appendix III

Table of Atomic Masses (Based on Carbon-12)

	Symbol	Atomic No.	Atomic Mass		Symbol	Atomic No.	Atomic Mass
Actinium	Ac	89	[227]*	Iridium	Ir	77	192.2
Aluminum	Al	13	26.9815	Iron	Fe	26	55.847
Americium	Am	95	[243]	Krypton	Kr	36	83.80
Antimony	Sb	51	121.75	Lanthanum	La	57	138.91
Argon	Ar	18	39.948	Lawrencium	Lw	103	[257]
Arsenic	As	33	74.9216	Lead	Pb	82	207.19
Astatine	At	85	[210]	Lithium	Li	3	6.939
Barium	Ba	56	137.34	Lutetium	Lu	71	174.97
Berkelium	Bk	97	[247]	Magnesium	Mg	12	24.312
Beryllium	Be	4	9.0122	Manganese	Mn	25	54.9380
Bismuth	Bi	83	208.980	Mendelevium	Md	101	[256]
Boron	B	5	10.811	Mercury	Hg	80	200.59
Bromine	Br	35	79.909	Molybdenum	Mo	42	95.94
Cadmium	Cd	48	112.40	Neodymium	Nd	60	144.24
Calcium	Ca	20	40.08	Neon	Ne	10	20.183
Californium	Cf	98	[249]	Neptunium	Np	93	[237]
Carbon	C	6	12.01115	Nickel	Ni	28	58.71
Cerium	Ce	58	140.12	Niobium	Nb	41	92.906
Cesium	Cs	55	132.905	Nitrogen	N	7	14.0067
Chlorine	Cl	17	35.453	Nobelium	No	102	[253]
Chromium	Cr	24	51.996	Osmium	Os	76	190.2
Cobalt	Co	27	58.9332	Oxygen	O	8	15.9994
Copper	Cu	29	63.546	Palladium	Pd	46	106.4
Curium	Cm	96	[247]	Phosphorus	P	15	30.9738
Dysprosium	Dy	66	162.50	Platinum	Pt	78	195.09
Einsteinium	Es	99	[254]	Plutonium	Pu	94	[242]
Erbium	Er	68	167.26	Polonium	Po	84	[210]
Europium	Eu	63	151.96	Potassium	K	19	39.102
Fermium	Fm	100	[253]	Praseodymium	Pr	59	140.907
Fluorine	F	9	18.9984	Promethium	Pm	61	[145]
Francium	Fr	87	[223]	Protactinium	Pa	91	[231]
Gadolinium	Gd	64	157.25	Radium	Ra	88	[226]
Gallium	Ga	31	69.72	Radon	Rn	86	[222]
Germanium	Ge	32	72.59	Rhenium	Re	75	186.2
Gold	Au	79	196.967	Rhodium	Rh	45	102.905
Hafnium	Hf	72	178.49	Rubidium	Rb	37	85.47
Helium	He	2	4.0026	Ruthenium	Ru	44	101.07
Holmium	Ho	67	164.930	Samarium	Sm	62	150.35
Hydrogen	H	1	1.00797	Scandium	Sc	21	44.956
Indium	In	49	114.82	Selenium	Se	34	78.96
Iodine	I	53	126.9044				

	Symbol	Atomic No.	Atomic Mass		Symbol	Atomic No.	Atomic Mass
Silicon	Si	14	28.086	Thulium	Tm	69	168.934
Silver	Ag	47	107.870	Tin	Sn	50	118.69
Sodium	Na	11	22.9898	Titanium	Ti	22	47.90
Strontium	Sr	38	87.62	Tungsten	W	74	183.85
Sulfur	S	16	32.064	Uranium	U	92	238.03
Tantalum	Ta	73	180.948	Vanadium	V	23	50.942
Technetium	Tc	43	[99]	Xenon	Xe	54	131.30
Tellurium	Te	52	127.60	Ytterbium	Yb	70	173.04
Terbium	Tb	65	158.924	Yttrium	Y	39	88.905
Thallium	Tl	81	204.37	Zinc	Zn	30	65.37
Thorium	Th	90	232.038	Zirconium	Zr	40	91.22

Appendix IV

Making Measurements—
Laboratory Techniques

For centuries people who made measurements set up systems of units to describe length, area, volume, and mass. These units often had rather romantic names that were created for a specific purpose. There was a myriad of such names, and among them we might mention:

acre	barrel	bolt	bushel
carat	chain	cord	cubit
dram	ell	em	fathom
furlong	gill	grain	hand
league	noggin	pace	perch
quire	rod	stone	ream

These days many of these units are still used, often in connection with only one application, such as horse racing (furlongs), a man's weight (stones), or printing (ems).

Scientists early on realized that it would be advantageous to have one system of measurement, to be used by all, and several were suggested, and used by your authors when they were young students. These were mostly based on the metric approach, with different sizes related one to another by factors of 10. Thus we have the meter, equal to 100 centimeters, or 1000 millimeters. In 1960, a group of scientists set up the International System of Units, or SI, a coherent system of seven base units, which can be used to describe most measurements one is likely to make, and which can be used together in many natural laws. The five units we will use are the:

meter	kilogram	second	Kelvin	mole

Although in principle one might expect to use these units directly in reporting data, some of them turn out to have awkward magnitudes in many situations. So we use SI units when it is convenient, but for the first two we tend to work more with related metric units, or with unrelated, but physically realistic, units. We will discuss our approach in the following sections. (Perhaps not surprisingly, none of the units in our first list are SI, or metrically related to SI.)

Mass

One of the most important measurements you will be making involves determining the mass of a sample. The unit we will almost always use is the gram, rather than the kilogram, since a kilogram is just too large, about two pounds. The gram is not a base unit in SI, but is easily converted to the kilogram when necessary.

Measurement of mass is almost always done by determining by one means or another the force exerted on a sample by gravity. This force is proportional to the mass of the sample. The device that is used is called a balance, because the mass of the sample is established by balancing it against a standard mass or force. In Figure IV.1 are shown several balances that were used in the past century. The balance on the left was used until about 1950. Using it required a set of weights that were put on one pan until the two pans came to the same height; with care one could weigh a sample to 0.0001 grams, but it would take several minutes to make one weighing. There were a lot of drawbacks to such a balance, but they were used for over a century in the form shown or more primitive ones. After World War II, instrument design developed rapidly. The balance in the middle is a mechanical one, but the weights are internal, and added to or taken from a pan inside the balance. Final readings are taken from a set of dials that add or take off weights, and an optically projected

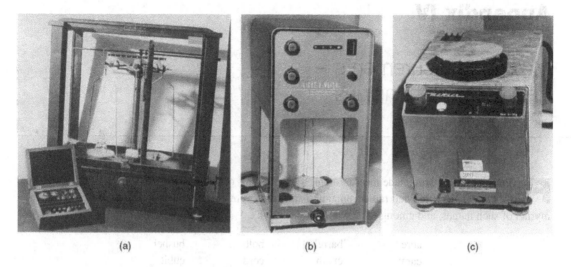

(a)   (b)   (c)

Figure IV.1   (a) An old mechanical analytical balance circa 1940—good to 0.0001 g, maybe. (b) A modern mechanical analytical balance circa 1960—maximum load 120 g, good to 0.0001 g. (c) A modern top-loading semi-automatic balance circa 1970—good to about 1 mg. *(Sherman Schultz)*

scale that furnishes masses in the milligram range. This type of balance is still used in some schools, and if you have one in your laboratory, your instructor will show you how it operates. The balance on the right is called a top-loading balance; mass is read directly from dials and an optical scale; this kind of balance gives rapid results but is limited in its precision to milligrams at best.

In recent years electronic balances have been developed that furnish the mass directly with a digital readout (see Figure IV.2). They do not have any weights, but depend on balancing the sample mass by a magnetic force that can be accurately related to mass. These balances can be accurate to 0.0001 g, and may arrive at the mass of a sample within 30 seconds or less. In a very real sense, they are the ultimate weighing machines. In using such a balance, you first depress the control bar. This will zero the balance, and it will read 0.0000 g. Place your sample on the balance pan, close the balance door, and read the mass when the balance gives a steady reading. If your sample is to be in a container, you can find its mass by weighing the empty container, then depressing the control bar to re-zero, or tare, the balance. Then when you add your sample, its mass will appear directly as the digital readout. There are many brands of these balances, so if your lab has one, your instructor will describe the details of the operation of your balance before you use it.

Here are some guidelines for successful balance operation, which apply to the weighing of a sample on any balance:

1. Be certain the balance has been "zeroed" (reads 0 grams) before you place anything on the balance pan.
2. Never weigh chemicals directly on a balance pan. Always use a suitable container or weighing paper. In several experiments you will weigh a sample tube and measure the mass of sample removed by difference.
3. Be certain that air currents are not disturbing the balance pan. In the case of an analytical balance, always shut the balance case doors when making measurements you are recording.
4. Never put hot, or even warm, objects on the balance pan. The temperature difference will change the density of the air surrounding the balance, and thus give inaccurate measurements.
5. Record your measurement, to the proper number of significant figures, directly on to your Data page or in your notebook.
6. After finishing your measurements, be certain that the balance registers zero again and close the balance doors. Brush out any chemicals which may have been spilled.
7. Be gentle with your balance. It is a sensitive, rather delicate, instrument, and, like a person, responds best when treated properly.

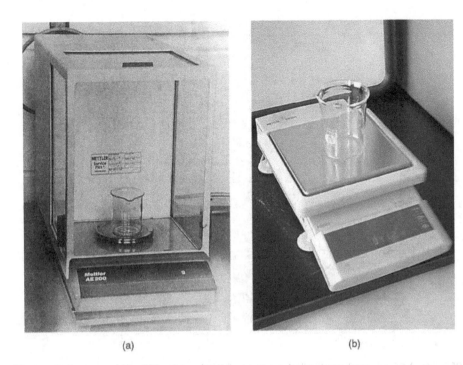

(a)   (b)

Figure IV.2   (a) A modern electronic automatic analytical balance, circa 1985. Its maximum load is 120 g, and it is good to 0.0001 g. The beaker weighs 27.5056 g. (b) An electronic top-loading balance with a max load of 3.6 kg and good to 0.1 g. *(Mary Lou Wolsey)*

Volume

Another very common measurement we make in the laboratory is that of volume. The base SI unit of volume is the cubic meter (m^3). Again, this is much much larger than you ordinarily encounter. A large bathtub might contain a cubic meter of water, roughly 250 gallons. The units of volume we ordinarily use are the liter, L, and the milliliter, mL. A liter equals 0.001 m^3, and a mL equals, 1×10^{-6} m^3, so both are derived from the SI unit.

The cubic cm, sometimes called the cc, pronounced "see see", is equal in volume to the mL, and is a term often used in hospitals.

$$1 \text{ mL} = 1 \text{ cc} = 1 \text{ cm}^3$$

If we are asked to add 150 mL of water to a beaker during an experiment in which we are carrying out a chemical reaction, we don't need high accuracy, and might use a beaker or flask on which there are some volume markings (as shown in Figure IV.3), which are good to about 5%. Somewhat more precise volumes are obtained with a graduated cylinder, where one can get within about 1% of a needed volume.

A. Pipets

More precise volume measurements are often required; these may be obtained with pipets, which are available in many different volumes, from 1 mL up to about 100 mL. In several experiments in this manual pipets are used. Pipets are calibrated to deliver the specified volume when the meniscus at the liquid level is coincident with the horizontal line etched around the upper pipet stem. The following steps make for proper use of a pipet:

1. The pipet must be clean. When it drains, there should be no drops left on the pipet wall. It does not need to be dry.
2. Always use a pipet bulb, not your mouth, when pulling liquid into a pipet. You shouldn't run the risk of getting the liquid into your mouth.
3. Place the pipet bulb on the upper end of the pipet and squeeze the air out. Immerse the tip of the pipet into the liquid and draw up enough liquid to get some into the main body of the pipet. Remove the bulb and

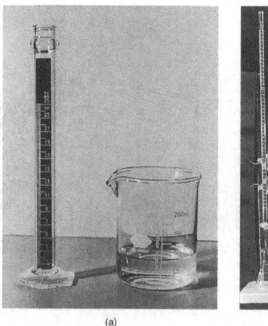
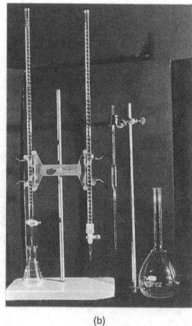

(a)   (b)

Figure IV.3   (a) A 25-mL graduated cylinder, good to 0.2 mL, and a 250-mL beaker with graduations, good to 20 mL. (b) A pair of 50-mL burets, a 10-mL pipet, and a 1000-mL volumetric flask. The burets and pipet are good to 0.01 mL, and the flask is good to about 0.1 mL. *(Sherman Schultz)*

place your finger over the top of the pipet stem. Hold the pipet in the horizontal position and swirl the liquid around inside, rinsing the upper stem and body. Drain the liquid into a beaker; repeat the rinsing twice with small volumes of the liquid. This ensures that the contents of the pipet will be the liquid you wish to work with, and not water or a previously-used reagent. Finally, using the bulb, fill the pipet, drawing liquid a centimeter or two above the calibration mark. Put your finger over the top of the stem and, carefully, let liquid flow out into a beaker until the bottom of the meniscus is just at the level of the calibration mark. Wipe off the lower end of the pipet with a tissue and place the tip inside the receiving container, with the tip touching the wall. Release the liquid and let the liquid flow into the container. Hold the tip against the wall for about 30 seconds after it appears that all the liquid has been transferred, to make sure all the liquid drains out.

It takes skill and practice to make good use of a pipet. It is now possible to obtain pipettors, which automatically and repeatedly deliver precise volumes of reagents. These speed up the pipetting procedure enormously, and you may have them in your laboratory.

B. Burets

A buret is a long calibrated tube fitted with a stopcock to control release of reagent liquids. Burets are used in a procedure called a titration, and are frequently used in pairs. This procedure allows one to add precisely measured volumes of reagents to a so-called "end point," at which the amount of one reagent is equivalent chemically to another, which may also be a liquid in a buret, or a weighed solid. We have several titration experiments in this manual (see Fig. IV.3b).

As with a pipet, working with a buret requires skill and practice. The following procedure should be helpful:

1. Check the buret for cleanliness by filling it with distilled water and allowing it to drain out. A clean buret will leave an unbroken film of water on the interior walls, with no greasy beads. If necessary, clean your buret with soapy water and a long buret brush. Rinse with tap water, and finally with distilled water.

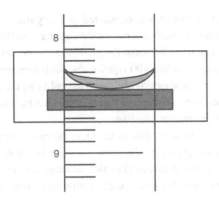

Figure IV.4   Reading a buret. The volume reading is 8.46 mL.

2. Rinse the buret three times with a few milliliters of the reagent to be used. Tip the buret to wash the walls with the reagent, and drain through the tip into a beaker. With the stopcock closed, fill the buret with reagent to a level a little above the top graduation, making sure to fill the tip completely. Open the stopcock carefully and let the liquid level go just below the top zero line. Read the level to the nearest 0.01 mL. To do this, place a white card with a sharp black rectangle on it (a piece of black tape is ideal) behind the buret, so that the reflection of the bottom of the meniscus is just above the upper black line, and at eye level (see Fig. IV.4). Then make the volume reading.

   Add the reagent until you get to the end point of the titration, which is usually established by a color change of a chemical called an indicator. To hit the end point within one drop, you need to add reagent slowly when the titration is nearly complete, more rapidly at the beginning. Swirl the container to ensure the reactants are well-mixed. Often you get a clue from the indicator that you are near the end point. If you go past it, you can use the first titration as a guide for the second or third trial. Read the liquid level as before to 0.01 mL.

C. Volumetric Flasks

A volumetric flask is one which has been especially calibrated to hold a specified volume: 10.00 mL, 25.00 mL, etc. These flasks are used when accurate dilutions are required in analytical experiments. Place the solute or solution in the previously cleaned, but not necessarily dry, flask. Add water to bring the volume up to the mark on the flask. After the liquid volume reaches the lower neck of the flask, add the water with a wash bottle. The last volume increments should be added dropwise until the bottom of the meniscus is coincident with the mark on the flask. Stopper the flask and mix thoroughly by inverting 20 times.

D. Pycnometers

The most precise volume measurements are made with a pycnometer, such as the one used in the first experiment. A pycnometer is simply a container with a very well-defined volume, such as a small flask fitted with a ground glass stopper. With such a device, and a liquid with an accurately known density, one can determine the volume of the pycnometer to 0.001 mL, which is much better than you can do with a pipet or buret. Knowing that volume, you can measure the density of an unknown liquid with an accuracy equal to that of the density of the calibrating liquid, usually within 0.01%.

Temperature

The SI unit of temperature is the Kelvin, which is equal in size to the °C. At 0°C, the Kelvin temperature is 273.15 K. The degree sign, °, is not used when reporting Kelvin temperatures.

   Measurement of temperature is always done with a thermometer. Your lab drawer probably contains a standard laboratory thermometer suitable for temperature readings between about −10°C and 100°C, with 1° graduations. We will use this thermometer in experiments where high accuracy is not needed.

The typical glass thermometer contains a bulb connected to a very fine capillary tube. It contains a liquid, usually alcohol or mercury, which fills the bulb and part of the capillary. As the temperature goes up, the liquid expands and the level rises in the capillary. To make the graduations, one would in principle find the level at 0°C and at 100°C, and divide that interval into 100 equal parts. With mercury, the scale fits the one obtained from ideal gas behavior quite well. With other liquids, you must calibrate the thermometer at several places if you are to obtain reliable temperatures. Your standard thermometer can be read to about 0.2°C if you are careful, but the actual error is likely to be greater than that.

In some experiments it is desirable to be able to get more precise temperature values than are possible with the standard thermometer. You may be furnished with a digital thermometer which has a wide range and reads directly to 0.1°C. The sensing tip on that kind of thermometer contains a thermistor, similar to a transistor, which has electrical resistance that varies markedly with temperature. In recent years these thermometers have become widely used.

Glass thermometers are fragile and relatively expensive, so be careful when using them. If you slip a split rubber stopper around the thermometer and support it with a clamp you will minimize breakage, and lab fees. If you break a mercury thermometer, contact your lab supervisor. Liquid mercury has a low vapor pressure, but that vapor is very toxic, so it is important to clean up mercury spills completely, which is not easy. With limited-range thermometers, do not heat them above their maximum readable temperature, since that can ultimately break the thermometer.

When making temperature readings, allow enough time for the level to become steady before noting the final temperature. This will probably take less than a minute, so if you check the reading over a period of time you should be able to get a reliable value.

To obtain temperatures with an accuracy better than 0.1°C is not a simple matter. Some mercury thermometers read to 0.01°C, but they have large bulbs and very fine capillaries and are very expensive, about $250 each. Still higher precision can be obtained using thermistors. By making careful resistance measurements, one can get temperatures accurate to about 0.001°C.

Pressure

In studying the behavior of gases one must be able to measure the gas pressure. In the chemistry lab this is done by comparing the pressure with that exerted by the atmosphere, using a device called a manometer. A manometer is a glass U-tube partially filled with a liquid, usually water or mercury.

The atmospheric pressure is measured with a mercury barometer. This consists of a straight glass tube about 80 cm long, which is initially completely filled with mercury. The tube is then inverted while the open end is immersed in a pool of mercury. The mercury level in the tube falls. The observed height of the Hg column above the pool, as read from a scale behind the tube, is equal to the pressure exerted by the atmosphere, and is called the barometric pressure. One can also measure the barometric pressure with an aneroid barometer, which consists of a spirally-wound, flexible-walled flat metal tube containing a fixed amount of air. As the external pressure changes, the tube flexes, which turns a needle that records the pressure.

Barometric pressure is often reported in mm Hg, but it may be given in other units, such as atmospheres, by comparing the value in mm Hg to 760 mm Hg, the standard atmospheric pressure. In SI the pressure is given in either Pascals or bars. One standard atmosphere equals 1.01325×10^5 Pascals, or 1.0325 bars. In this manual we will report gas pressures in mm Hg or mm H_2O, since these are most easily measured directly. We can convert pressures in mm Hg to mm H_2O, or vice versa, by using the conversion factor: 1 mm Hg = 13.57 mm H_2O.

Having found the barometric pressure, it is a simple matter to measure the gas pressure in a flask such as that shown in Figure IV.5. In Experiment 8, the pressure inside the flask is equal to the barometric pressure, in mm H_2O, plus the pressure exerted by a water column of length $h_2 - h_1$, the water levels in the right and left arms of the U-tube. If you know those heights in mm,

$$P_{gas} = (P_{bar} + h_2 - h_1) \text{ mm } H_2O$$

When reading the H_2O levels in a manometer, take the height at the bottom of the water meniscus in the two arms, which you can read to 1 mm or better.

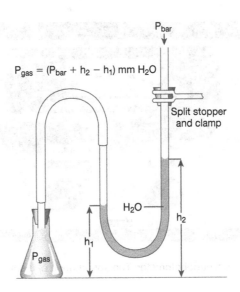

$$P_{gas} = (P_{bar} + h_2 - h_1) \text{ mm } H_2O$$

Figure IV.5 Measuring the pressure of a gas.

Light Absorption

In many chemical experiments the interaction of light with a sample can be used to furnish very useful information. In this laboratory course we will often determine the concentration of a species in a solution by measuring the amount of light at a given wavelength (color) that is absorbed by the sample. In an instrument called a spectrophotometer one can do this very easily. The spectrophotometer contains a device called a monochromator (usually made from a diffraction grating), which allows only one wavelength of light to pass through a sample. The absorbance, A, of the sample at that wavelength can be read directly from a meter on the instrument. By a famous equation called Beer's Law, the absorbance is proportional to the concentration, c, of the sample, usually the molarity:

$$A = K \times c$$

where K is a constant that depends on the sample, the container, and the wavelength of the light. Most samples obey Beer's Law. To find the concentration of a species in a sample, one measures the absorbance at an appropriate wavelength of a solution containing a known concentration of the species. From that absorbance, or several absorbances obtained with other concentrations, we can make a calibrating graph of concentration vs. absorbance and then use that graph to analyze unknown solutions containing that species. Your instructor will probably furnish you with such a graph in some of the experiments you perform.

A commonly used spectrophotometer is a Spectronic 20 (see Fig. IV.6). This instrument has two knobs on the front face. On the left is a zero transmittance adjustment knob and on the right a 100% transmittance adjustment knob. On the top surface at the left is a covered sample chamber, and at the right there is a wavelength selection knob. The Absorbance is read from the upper meter or display register.

To use the spectrophotometer, turn it on about 15 minutes before you need to operate it, to give the components time to warm up. Note the wavelength setting, which is one where the sample absorbs a moderate amount of the incident light. Since the sample is probably colored, the wavelength will most likely be in the visible region of the light spectrum. With the sample compartment closed, turn the zero adjustment knob until the needle or display register indicates 0% transmission. Open the sample chamber and insert a sample tube that is about 3/4 full of a reference solution (usually pure water). Shut the sample chamber and turn the 100% adjustment knob until the needle reads 100% transmission (this will be zero on the absorbance scale). Repeat the adjustment steps until no further changes are necessary. Fill the sample tube about 3/4 full of the solution being studied, insert it in the sample chamber, and close the chamber. Read the Absorbance from the meter or register. Then use the calibration graph to obtain the concentration of the species in the solution.

Figure IV.6   A Spectronic 20 spectrophotometer. The absorbance of a solution can be read to about 1%.
(Sherman Schultz)

pH Measurements

The pH of a solution is used to describe the concentration of the H^+ ion in that solution. To determine pH one can use acid-base indicators, which change color as the pH changes. Each indicator has a characteristic pH at which the color change occurs, so can be used to indicate whether the pH is higher, or lower, than the characteristic value. By employing several indicators one can fix the pH of a solution within a few tenths of a unit.

Most accurate pH measurements are made with a pH meter. The pH meter is a very high resistance voltmeter which measures the difference in voltage between a reference electrode (usually a so-called calomel electrode) and a glass indicating electrode whose potential is a function of the H^+ ion concentration in the solution in which it is immersed. The meter indicates the pH directly. The two electrodes may be separate, but more commonly are together in one combination electrode. The electrode is kept wet in an appropriate storage solution when not in use.

When you are asked to find the pH of a solution, you may first need to calibrate the pH meter, although it is likely that that was done by your lab supervisor prior to the laboratory session. To calibrate the meter you will need a buffer solution with a well-defined pH as a reference. Put about 25 mL of the buffer in a small beaker. Remove the electrode from the storage solution, rinse it with a stream of distilled water from your wash bottle, and then blot the electrode with a clean tissue. Place the electrode in the buffer. (The electrode is fragile, and expensive, so treat it gently.) Allow the system to equilibrate for a minute or two, or until the pH reading becomes steady. Then use the pH adjust knob to bring the displayed pH to that of the reference buffer solution.

Remove the electrode from the buffer, rinse it with distilled water as before, blot with a tissue, and place the electrode in the solution being studied. Allow time for equilibration, and record the pH. Rinse the electrode and return it to the storage solution.

Separation of Precipitates from Solution

A. Decantation

Decantation is used to remove a liquid from a precipitate by pouring it off. Allow the solution to set for a period of time until the precipitate is on the bottom of the beaker. Position your stirring rod across the beaker with one end protruding beyond the lip. With the index finger of one hand holding the rod in place, pour the liquid slowly down the stirring rod into a receiving vessel. Try not to disturb the precipitate as the last bits of liquid are poured off.

B. Filtration—The Buchner Funnel

Filtration is used when you wish to recover the precipitate in pure form. This is required in Experiment 3. You will use a Buchner funnel, which contains a piece of filter paper of appropriate size covering the holes at the base of the funnel. The Buchner funnel is connected by a rubber stopper or adaptor to the top of a side-arm filter flask. The side arm of the flask is connected via a series of rubber tubes and a safety trap to a water aspirator. When the water is turned on, a vacuum is created in the system by the water rushing by the side opening in the aspirator (see Fig. 3.2).

When you are ready to filter, moisten the filter paper with water, then turn the faucet on to start the vacuum. With the water faucet in the fully open position, pour the slurry of solution and precipitate down a stirring rod, as described under decantation. Wash out any remaining precipitate with a slow stream of liquid from your wash bottle. The procedure may call for washing the precipitate on the Buchner funnel with an appropriate volatile liquid before passing air through the filter cake for several minutes to complete drying. Turn off the water, disconnect the tube from the filter flasks, and remove the funnel.

C. Centrifugation

Centrifugation is used to aid in separation of a precipitate from a solution in a test tube. Using the centrifugal force generated by spinning the test tube at several hundred revolutions per minute gives an outward force for settling that is much higher than the gravitational force (see Fig. IV.7).

To use the centrifuge, be certain that the test tube containing the precipitate is not overly full—the liquid should be at least 3 cm below the top of the tube. Check to see that the test tube has no cracks, as these will cause the tube to break during centrifugation. Place the test tube containing your precipitate and solution in one of the centrifuge tubes. Place a blank tube containing the same amount of water as you have in your sample tube in the centrifuge tube that is opposite your sample tube. Turn on the centrifuge, allow it to spin for about a minute, and then turn it off. Keep your hands away from the spinning centrifuge top. After the spinning top has come to rest, remove the tube. Decant the supernatant liquid from the precipitate.

Qualitative Analysis—A Few Suggestions

In the course of your laboratory program you may do several experiments involving qualitative analysis. In those experiments we will use small test tubes and small beakers as sample containers. Separations of precipitates from solutions are always done by centrifugation.

In many procedures you will be told to add 1 mL, or 0.5 mL, to a mixture. This is best accomplished by first finding out what 1 mL volume looks like in a test tube. So, measure out 1 mL in a small graduated

cylinder, and pour that into the test tube, noting the height reached by the liquid in the tube. From then on, use that height to tell you the volume to add when you need 1 mL. Half that height indicates 0.5 mL.

Many of the steps in qualitative analysis involve heating a mixture in a boiling-water bath. To make the bath, fill a 250-mL beaker about 2/3 full of water. Support the beaker on a wire screen on an iron ring. Heat the water to the boiling point with your Bunsen burner, and then adjust the flame so as to keep the water at the simmering point. Use this bath to heat the test tube when that is called for. You can remove the tube with your test tube holder.

In separating a solid from a solution, we first centrifuge the mixture. The supernatant liquid is decanted into a test tube or discarded, depending on the procedure. The solid must then be washed free of any remaining liquid. To do this add the indicated wash liquid and stir with a glass stirring rod. Centrifuge again and discard the wash. Keep a set of stirring rods in a 400-mL beaker filled with distilled water. Return a rod to the beaker when you are done using it, and it will be ready when you next need it.

Pay attention to what you are doing, and don't just follow the directions as though you were making a cake. Try to keep in mind what happens to the various cations in each step, so that you won't end up pouring the material you want down the drain and keeping the trash. When you need to put a fraction aside while you are working on another part of the mixture, make sure you know where you put it by labeling the test tube holder with the number of the step in which you will return to analyze that fraction.

Appendix V

Mathematical Considerations— Making Graphs

In the laboratory you will be carrying out experiments of various sorts. Some of them will involve almost no mathematics. In others, you will need to make calculations based on your experimental results. The calculations are not difficult, but it is important you make them properly. In particular, you should not report a result that implies an accuracy that is greater, or smaller, than is consistent with the accuracy of your experimental data. In making such calculations we resort to the use of significant figures as a guide to proper reporting of our results. In the first part of this appendix we will discuss significant figures and how to use them.

In several experiments we will carry out a series of measurements on the same system under different conditions. For example, in Experiment 8 we measure the pressure of a sample of gas at different temperatures. In such experiments it is helpful to draw a graph representing our data, since it may reveal some properties of the sample that are not at all obvious if we just have a table of data. The second part of this appendix will show how to interpret a graph and how to construct one properly.

Significant Figures

In the laboratory we make many kinds of measurements. The precision of the measurement depends on the device we use to make it. Using an analytical balance, we can measure mass to ±0.0001 g, so if we have a sample weighing 2.4965 g, we have *five* meaningful figures in our result. These figures are called, reasonably enough, significant figures. If we measure the volume of a sample using a graduated cylinder, the volume we obtain depends on the cylinder we use. If we have a volume of about 6 mL and measure it with a 100-mL cylinder, we would have difficulty distinguishing between a volume of 6 mL and 7 mL, and would be unable to say more than that the volume was about 6 mL. That volume contains only *one* significant figure. With a 10-mL cylinder, we could measure the volume more precisely, report a volume of 6.4 mL, and have confidence that both figures in our result were meaningful, and that it contained *two* significant figures. Many measurements, perhaps most, can be interpreted as we have here.

Sometimes it is not clear how many significant figures there are in a number. Say, for example, we are told to add 1 mL of liquid from a pipet to a test tube. A pipet is a precisely made device, and can deliver 1.00 mL of liquid if used properly. In such a case, it would be sensible to say that there are really three significant figures in that volume, rather than one. It would probably have been better to be told to add 1.00 mL with a pipet, but often that is just not done, even by the authors of this manual.

Some numbers are exact and have no inherent error. There will be an integral number, like 12, or 19, students in your lab group. There is no way there will be 14.5 students, unless there is something strange going on. Conversion factors, used to convert one set of units to another, often contain exact numbers: 1 meter = 100 cm. Both numbers are exact, with no error at all. With volumes, 1 liter = 1000 mL. Again, both numbers are exact. If we convert from one system to another, then usually only one of the numbers in the conversion factor is exact: 1 mole = 6.022×10^{23} molecules. Here the 1 is exact, but the second number is not, and has four significant figures.

If you are in doubt as to the number of significant figures in a given number, there is a method for finding out that usually works. We write the number in exponential notation, expressing the number as the product of a number between 1 and 10, times a power of ten.

$$26.042 = 2.6042 \times 10^1 \quad 0.0091 = 9.1 \times 10^3 \quad 605.20 = 6.0520 \times 10^2$$

The first number in the product has the number of significant figures we seek. So, the first of the above numbers has 5 significant figures, the second has 2, and the third has 5 (a trailing zero is significant). There is one problem with this approach, and that is that sometimes we are given a number with no decimal point, like 2400. We cannot be sure how many significant figures are in that number, since, if the number were rough, there might be 2, but there could be 4 or even more. Here, you have to use some judgment, but in the absence of any guide, you would choose 4.

Students usually have little trouble deciding on the number of significant figures in a piece of data. The difficulty arises when the piece of data is used in a calculation, and they have to express the result using the proper number of significant figures. Say, for example, you are weighing a sample in a beaker and you find that:

$$\begin{aligned} \text{Mass of sample plus beaker} &= 25.4329 \text{ g} \\ \text{Mass of beaker} &= 24.6263 \text{ g} \\ \text{Mass of sample} &= 0.8066 \text{ g} \end{aligned}$$

Even though the two measured masses contain 6 significant figures, the mass of the sample contains only 4. In another case, where we mix some components to make a solution, weighing each of the components separately on different balances, we get:

$$\begin{aligned} \text{Mass of beaker} &= 25.5329 \text{ g} \\ \text{Mass of water} &= 14.0 \text{ g} \\ \text{Mass of salt} &= 6.42 \text{ g} \\ \text{Total mass} &= ? \end{aligned}$$

Here we have several pieces of data, each with a different degree of precision. The first mass is good to 0.0001 g, the second to 0.1 g, and the third to 0.01 g. If we take the sum, we get 45.9529 g, but we certainly can't report that value, since it could be off, not by 0.0001 g, but by 0.1 g, the possible error in the mass of the water. We can't improve the quality of a result by a calculation, so, using good sense, we can only report the mass to ±0.1 g. Since the measured mass is closer to 46.0 g than to 45.9 g, we round up to 46.0 as the reported mass, and have a mass with 3 significant figures. Generalizing, *in adding or subtracting numbers, round off the result so that it has the same number of decimal places as there are in the measured quantity with the smallest number of decimal places*. Round the last digit up if the number that follows is greater than 5, and don't round up if it is not.

When *multiplying* or *dividing* measured quantities, the rule is quite simple: *The number of significant figures in the result is equal to the number of significant figures in the quantity with the smallest number of significant figures.*

In the first experiment we measure the density of a liquid using a pycnometer, which is simply a flask with a well-defined volume. We find that volume by weighing the empty pycnometer and then weighing the pycnometer when it is full of water. We might obtain the following data:

$$\begin{aligned} \text{Mass of pycnometer plus water} &\quad 60.8867 \text{ g} \\ \text{Mass of empty pycnometer} &\quad 31.9342 \text{ g} \\ \text{Mass of water} &\quad 28.9525 \text{ g} \end{aligned}$$

We are asked to find the volume of the pycnometer, given that the density of water is 0.9973 g/mL.

$$\text{density} = \frac{\text{mass}}{\text{volume}} \quad \text{so} \quad \text{volume} = \frac{\text{mass}}{\text{density}} = \frac{28.9525 \text{ g}}{0.9973 \text{ g/mL}} = 29.030883 \text{ mL}$$

The mass contains 6 significant figures, the density has 4, and the results from a calculator have 8 or more. Since the density has the smallest number of significant figures, four, we report the volume to 4 significant figures, as 29.03 mL.

In most cases, the reasoning that is required to properly report a calculated value of a property is no more complicated than that we used here. Remember, you can't improve the accuracy in a result by means of a calculation. If you note the quality of your data, you should be able to report your result with the same quality. As with most activities, a little practice helps, and so does a little good sense.

Graphing Techniques

In many chemical experiments we find that one of our measured quantities is dependent on another. If we change one, we change the other. In such an experiment we find data under several sets of conditions, with each data point associated with two measured quantities. In such cases, we can represent these points on a two-dimensional graph, which shows in a continuous way how one quantity depends on another.

Interpreting a Graph

The first graph in the lab manual is in Experiment 3, and is a graph of this kind. It is shown in Figure V.1. The graph describes how the solubility of each of two compounds, KNO_3 and $CuSO_4 \cdot 5 H_2O$ depend on temperature. The temperature is shown along the horizontal axis, called the abscissa, over the range from 0 to 100°C. The solubility, in grams per 100 grams water, is along the vertical axis, sometimes called the ordinate. The x's along each of the two curved lines in the graph are data points, showing measured solubility at given temperatures.

The graph has some implied features that are not shown. We actually made the graph on a piece of graph paper on which there was a grid of parallel lines. On the vertical lines the temperature is constant. If you draw a vertical line upwards from the 50°C hashmark on the vertical axis, the temperature on that line is always 50°C. On the vertical line at 73°C, three-tenths of the distance between the 70 and 80 hashmarks, it is always 73°C, and so on. Along horizontal lines the solubility is constant. The horizontal line drawn through the 80 hashmark represents a solubility of 80 grams in 100 grams of water all along its length. When we obtained the data for the solubility of KNO_3, we found that, at 60°C, we could dissolve 105 g KNO_3 in 100 g of water. We entered the data point we show as A on the graph at 60°C and 105 g KNO_3. The rest of the data points for the two compounds were entered in the same way. We then drew a smooth curve through those points, and took out the grid.

Given a graph like that we have shown, you can extract the data by essentially working backwards. To find the solubility of KNO_3 at 60°C, draw the dashed line up from 60°C. The solubility is given by the point at which that line intersects the KNO_3 graph line. Draw a horizontal line from that intersection to the vertical axis and read off the solubility of KNO_3, which you can see is about 105 g/100 g H_2O. If you wanted to find the amount of water you would need to dissolve 21 g KNO_3 at 100°C, you would draw the vertical line at

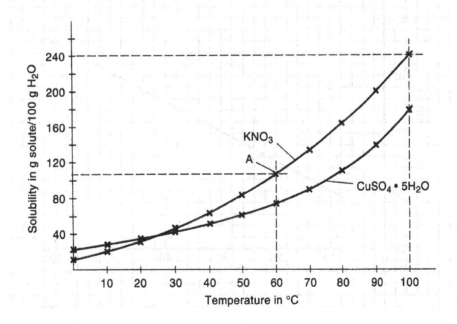

100°C up to the KNO$_3$ line; then draw the horizontal line at the intersection with the KNO$_3$ curve, and you find that at 100°C, about 240 g KNO$_3$ will dissolve in 100 g H$_2$O. Then a simple conversion factor calculation would give you the amount of water needed to dissolve 21 g KNO$_3$. Clearly, there is a lot of information in Figure V.1 if you interpret it completely. Its meaning is almost, but not quite, obvious.

Making a Graph

Now let us construct a graph from some experimental data. Given the grid in Figure V.2, we measure the pressure of a sample of air as a function of temperature. One of our students obtained the following set of data:

| Pressure in mm Hg | 674 | 739 | 784 | 821 |
| Temperature in °C | 0.0 | 24.5 | 42.0 | 60.1 |

To make the graph we need to decide what goes where. Since the pressure is dependent on the temperature, we put the pressure on the vertical axis, and the temperature on the horizontal axis, labeling those axes. The temperature goes from 0°C to 60°C; we want the graph to fill the grid, not just be in one corner, so we select a temperature interval between grid lines that will fill the grid. Since we have about 30 vertical grid lines available, and need to cover a 60°C interval, we put the 10° hashmarks at intervals of five grid lines, with 2° between grid lines. Since there are about 20 horizontal grid lines, and we need to cover a pressure change of about 150 mm Hg, we make the lowest pressure 650 mm Hg and the interval between grid lines equal to 10 mm Hg. We put in the hashmarks at 50-mm intervals. Having laid out the graph, we insert each of the data points, using x's or dots. The points fall on a nearly straight line, so we draw such a line, minimizing as best we can the sum of the distances from the data points to the line. If we wish, we can find the equation for that line, which, since it is straight, is of the form y = mx + b where m is the slope of the line and b is the value of y when x equals zero. Using Excel, our student found that the equation was y = 2.466x + 676.4. That equation can be used to find another useful piece of information about the properties of gases, but we need not go into that here.

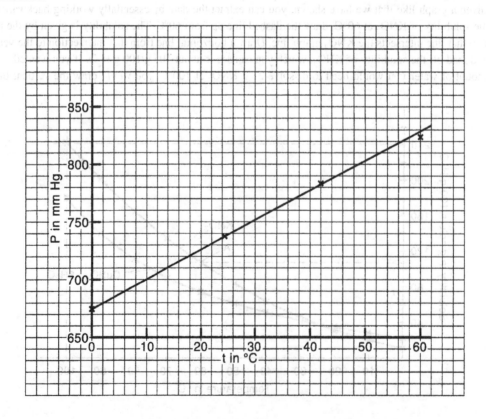

The key to successful construction of graphs is to properly select the variables and assign them to the two axes. Label those axes. Then, given the changes in the variables, select grid intervals that will use up the area of the grid. Put in the hashmarks for each variable at appropriate intervals and label them with the values of the variable. Take the data points and enter them on the grid. If the line is theoretically straight, draw a straight line through the data points in such a way as to minimize the sum of the distances from the points to the line.

Many natural laws are not as simple as the one relating gas pressure to temperature. However, by a few rather easy tricks, involving choice of variables, one can obtain data points that will lie on a straight line. For example, in Experiment 15, we find that the vapor pressure of a liquid depends on temperature according to the following equation:

$$\ln VP = \frac{-\Delta H_{vap}}{RT} + C$$

where ln VP is the logarithm of the vapor pressure to the base e, T is the Kelvin temperature, and ΔH_{vap} and R are constants and equal to the heat of vaporization and the gas constant, respectively. The equation looks formidable and is. However, if we let x equal ln VP and y = 1/T, then the equation takes the form y = mx + b, a straight line with a slope m equal to $-\Delta H_{vap}/R$. By using those variables, we can find the slope of the straight line and determine ΔH_{vap} for the liquid. It is very possible that the equation was initially found by an imaginative scientist who measured the vapor pressure against temperature and tried plotting the data against different variables, seeking a straight line dependence. From such efforts many great discoveries have been made.

Appendix VI

Suggested Locker Equipment

2 beakers, 30 or 50 mL
2 beakers, 100 mL
2 beakers, 250 mL
2 beakers, 400 mL
1 beaker, 600 cc
2 Erlenmeyer flasks, 25 or 50 mL
2 Erlenmeyer flasks, 125 mL
2 Erlenmeyer flasks, 250 mL
1 graduated cylinder, 10 mL
1 graduated cylinder, 25 or 50 mL
1 funnel, long or short stem
1 thermometer
2 watch glasses, 75, 90, or 100 mm
1 crucible and cover, size #0
1 evaporating dish, small

2 medicine droppers
2 regular test tubes, 18×150 mm
8 small test tubes, 13×100 mm
4 micro test tubes, 10×75 mm
1 test tube brush
1 file
1 spatula
1 test tube holder, wire
1 test tube rack
1 tongs
1 sponge
1 towel
1 plastic wash bottle
1 casserole, small

Appendix VII

Introduction to Excel

Over the last several years a piece of software called Excel has become available for use in processing scientific data obtained in the laboratory, or, indeed, data involved in business activities. Essentially Excel is a spreadsheet, containing many cells in which data can be entered and manipulated. It is particularly useful when there are several, or many, experimental values for a property, such as the volume of a gas obtained at different temperatures under constant pressure. In this introduction we will show how to use Excel in some of its simpler applications. You will find it easy to use, but it does require some getting used to. To get started, you will need to have a computer on which the software has been installed. There are many updates of the program, but we will try to avoid references specific to any particular version.

Basic Layout

Click on the Excel icon on your computer. If there is none, your instructor will tell you how to call it up. You should see a screen, called a Workbook, on which there is a grid containing many rectangular boxes, called cells. Each cell has an associated label, with a letter indicating its column, and a number indicating its row. If you click on cell B5, it will be highlighted. If you put the cursor in that cell, type a number or a word, and click on enter or return on the keyboard, that data will appear in the cell. You can change the width of the cell by clicking and holding on the grid line in question, up where it intersects with the rectangle containing a letter.

Setting up Columns of Data

Let's proceed by entering the words "Temperature (C)" in Cell A1. Change the cell width if necessary to make the title fit. In Cell A3, type in the number 20 and press enter or return. The highlighted cell will automatically become A4. Enter the formula "= A3 + 20" in that cell, click on enter, and 40 should appear in A4. Highlight the A4 cell with the mouse, and set the cursor at the lower right-hand corner of that cell. You will find that the cursor will change from a white plus symbol to a solid black plus symbol or a square. While the cursor is a black plus or a square, click and drag the cursor down to select all the cells you want to include with that formula. When you release the mouse button, the numbers 60, 80, 100 . . . will appear in the column. This method is very useful if you have a column of data which are related by a formula.

Now enter the word "Temperature (K)" in Cell B1. In Cell B3, enter the formula "= A3 + 273", press enter, and 293 will appear in Cell B3. In Cell B4, type the formula "= A4 + 273". When you click on enter, the number 313 appears in that cell. Using the same approach we used in Column A, find the Kelvin temperatures in Column B for each of the Celsius temperatures in Column A.

In Cell C1, type the word "Volume (L)". Let's use the Ideal Gas Law, $PV = nRT$, to find the volume in liters of a mole of gas at one atmosphere pressure at the Celsius temperatures in Column A. By the Gas Law, $V = nRT/P$; in our example, $P = 1$ atm, $R = 0.0821$ L-atm/moleK, and T is the Kelvin temperature. So in Column C, V equals 1*0.0821*T, or 0.0821*T. (The mathematical signs for operations in Excel do not correspond exactly to those you learned in school. + and − mean add and subtract, * means multiply, / means divide, and ^ means raise to a power. We write 2^3 as 2^3 in Excel. A number like $3.45*10^5$ must be written as 3.45E5.)

To find the values of Volume in Column C, first enter the number "0.0821" in Cell C10. In Cell C3, write the formula "= C10*B3", and press enter. In Cell C3 the number 24.055 should appear. Highlight Cell C3, put the cursor on the lower right-hand corner of the cell, and when the black plus symbol appears, drag down to the last row of data and release. The volumes in liters of gas at the various temperatures should appear in Column C.

You may wonder why we use \$C\$10 instead of just C10. Try C10 and see what happens. When you drag down the cursor to copy this cell, the program will increment not only C3 to C4, but also C10 to C11, and zeroes will appear in Column C, since C11 is zero. In many formulas, you will find that you wish to have a constant retain its value. You can do that by surrounding the letter in the name of the cell containing the constant with \$ signs.

Some Comments on Excel

In Excel, the program will assign the contents to a cell to one of three categories, a number, text material, or a formula. Numbers and text cause no trouble. Formulas may. You can write the formula describing how to evaluate the number that is to go into a cell by writing an = sign at the left end of the cell, or at the left end of the Formula Bar at the top of the screen, just above the column letter B, followed by an expression involving the contents of one or more cells, all of which must be known. The = sign must be at the far left, with no space to its left. The equation for the formula may be complex, but it must include the Excel symbols for any operation to be carried out. All referrals to particular cells must be to cells with known numerical contents. Parentheses should be carefully used to make it clear what you wish to have done.

If you make an error, and errors are inevitable, you may find nothing happens when you hit enter, or you may get one of several error messages, which may or may not be helpful. If you get into trouble, the delete or esc button may clear the screen to its form before the error occurred. Do not be reluctant to seek help, either from your lab supervisor or a fellow student more familiar with Excel than you are. Working with Excel can be frustrating, but sooner or later you will able to use it effectively.

Using Excel to Make Graphs

In some cases it is helpful to create graphs that show the relationships between experimentally measured variables.

Given two columns of data for two variables, like temperature and volume, on the Excel spreadsheet, obtained under different conditions, we can construct the graph that shows how the variables are related.

Before you proceed, click on View on the top line of the screen, and select Normal from the list.

Click on the Chart Wizard icon on the top row of Excel, about the third icon from the right.

This displays the Step 1-Chart Type, a screen showing several possible graph forms. Click on XY Scatter, which highlights a graph with scattered data points.

Click on Next, for Step 2-Chart Source Data. Click on Series. Click on Add and/or Remove, if necessary, to get one numbered Series in the Series box. On the screen that appears, click on Scale, if necessary, to bring down three rectangular boxes, Name, X Values, and Y Values. Delete the contents of any boxes that are not already empty; first empty X, and then Y. The cursor may insist on being in the Y box.

To get the Y values, put the cursor at the left end of the Y Values Box, and drag the mouse down the column of data containing the dependent variable, the Volume, in Column C. When you lift the mouse button, the information in that column will appear in the Y Values box.

To get the X values, put the cursor at the end of the X Values Box, and drag the mouse down the column of data containing the independent variable, the Kelvin temperature, in Column B. The information in that column will appear in the X Values box. A graph showing the relation between the variables will appear on the screen. The scale will probably be incorrect.

Click on Next, for Step 3. Click on Titles. Enter Volume vs Temperature as the title of the graph in the Title Box. Enter the name of the X-variable, Temperature (K), in the Value (X) axis Box. Enter the name of the Y-variable, Volume (L), in the Value (Y) axis Box. Click on Gridlines. Add or remove them as seems reasonable.

Click on Next, for Step 4. Decide where to place the graph, either on a separate page or as an Object on the spreadsheet. Usually, you'll want to check the latter.

Click on Finish, to bring down the final graph, labeled and titled, but probably not scaled as you would like. To fix the scale, click on the line of the axis you wish to change, which puts small squares on ends of that line. Click on the X-axis twice, which brings up the Format Axis screen. If necessary, click on Scale on that screen to bring up a table with boxes for min and max values of x that would be appropriate (250 and 400). Fill in those boxes, moving the graph if necessary to make it possible to see the x-data. (To move the graph, drag the cursor on the corner of the outer graph boundary, when the little squares are there.) Then, click on the Y-axis twice, click on Scale, and fill in the min and max values of the y-data (20 and 35). Enter those values, and the graph should be complete. Reclick on the X and Y axes as needed to readjust the scales so that the data points fill the graph nicely. If you would like to join the points by a line, click on Chart, so that the word "Chart" appears on the top line of the screen. Click on Chart, and then select the Add Trendline entry. (Press Options on the Trendline screen, and click on the Display Equation entry.) If necessary, select the Series on which the graph is based. Click on OK, and the trendline and its equation should appear on the graph.

In the process of doing these steps it is very easy to hit a wrong key or make another kind of error. This can cause many problems, which are not all easy to fix. If you find yourself in that pickle, start over by hitting backspace/delete, esc, or the extra del button on the right side of the keyboard. That should bring you back to the original data on the spreadsheet, and you can start over.